Queen Rearing Simplified

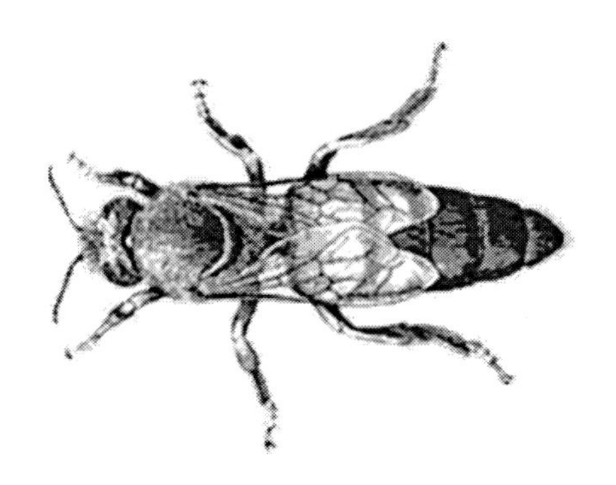

Jay Smith

Queen Rearing Simplified by Jay Smith

Originally Published by
THE A.I. ROOT COMPANY Medina, Ohio 1923

Reprinted 2011 by

X-Star Publishing Company
Nehawka, Nebraska, USA

ISBN: 978-161476-052-8

198 pages
63 illustrations

Transcriber's preface.

Queen Rearing Simplified is one of the most popular queen rearing books of all time written by a man who raised a lot of good queens. This is a good book on grafting. It is no longer in print, so I am trying to keep Jay's wisdom alive here. There are many queen breeding books by scientists or small-scale breeders, but this is by a beekeeper who raised thousands of queens every year. I think that is much more applicable to practical queen rearing. This is a reprint so the old pictures are not the highest quality. Many of the opinions stated here are outdated in Jay's point of view by the time of Jay's second book, *Better Queens* written a quarter century after this one. Better Queens is a graftless method. You can buy *his other book,* from X-Star Publishing Company at any online bookseller or you can read it online for free at **http://bushfarms.com/beesbetterqueens.htm**

This book is dedicated to my full partner - my wife. –Jay Smith

Introduction

For several years past there has been a growing interest in Queen-rearing, as more beekeepers are coming to recognize the important part the queen plays in beekeeping. I have been receiving a large amount of correspondence on the subject of Queen-rearing from beekeepers wishing for detailed information on the subject. Their many questions have prompted me to attempt this book, and to explain such points as are not clear to those interested in Queen-rearing.

To assist the honey producer in rearing his own queens is my primary object; but I also describe methods adapted to the amateur as well as the commercial queen-breeder. To the beginner in beekeeping, however I would recommend a careful study of one or more of the following books on general beekeeping before taking up this work; "The A B C and X Y Z of Bee Culture" (Root), "Beekeeping" (Phillips), "Starting Right with Bees" (Rowe), "Langstroth on the Honey Bee" (Dadant), "Fifty, Years Among the Bees" (Miller). In addition to bee books one should read all articles in the bee journals by able writers and especially those written by Geo. S Demuth, who is now generally recognized as our highest authority on beekeeping. For a description of different methods of Queen-rearing read Pellet's "Practical Queen Rearing."

I presenting this volume to the beekeeping public, nothing radically new or revolutionary is offered. The system described has been taken from many sources, so it is impossible to give credit to all who have contributed through their books and their writings to our bee journals.

More is due to Mr. G.M. Doolittle than any other, for to him we owe the invention of artificial cell cups and the art of grafting.

I shall deal mainly with the successes I have had and not with the failures. I have two reasons for doing this. One is that almost any beekeeper has failures without having to refer to a text-book on the subject; and the second is, that I wish to keep this book within modest dimensions. If I should chronicle all of my failures, a book so voluminous would result that a Webster's Unabridged might look like a vest-pocket edition in comparison.

The object of this book then, is not to present many new methods but to place before the reader, with the aid of the camera, such methods with variations as I have used for twenty-one years, and to describe them in detail so that any one wishing to rear queens can succeed, and, if failure comes, he may refer to this book, and find the cause of it. Many have reported indifferent success with the grafting method of queen-rearing. Upon investigation, it was frequently found they had followed all of the rules laid down with *one* or *two* exceptions. These very exceptions brought the failure. I hope this book may be of help to such. Frankly, I do not know whether it will or not. The reader must be the judge. Again, if this little volume interests some overworked business or professional man or woman, and, through it, pleasure and recreation are gained, and he is thus better able to meet some of the harsher things of life, I shall consider my efforts have not been in vain.

Vincennes, Indiana, October 5, 1923. JAY SMITH

Table of Contents

Chapter I. Importance of Good Queens.

In view of what has been said by the writers in the past, it would hardly seem necessary, if the best results are to be obtained in honey production, to call attention to the importance of having every colony headed by a good, prolific Italian queen.

You will note that I say a good *Italian* queen. Beekeepers are practically unanimous in the opinion that the Italian bees are much superior to Blacks in nearly all respects. They are better workers, swarm less, are more gentle and are much superior in cleaning out European foul brood. Unfortunately the black bee was introduced into the United States over two hundred years before the Italian, and therefore the Blacks have become pretty well established in all parts of our country. They are now found wild in trees and rocks in every state from coast to coast, and in many parts of Canada. Consequently, one very good reason why the honey producer should rear his own queens is to get rid of the black bees and hybrids.

Every beekeeper concedes the point that each colony must be headed by a good prolific queen, and all writers on the subject have emphasized it in the strongest terms, yet in truth very few of us fully realize the importance of good queens.

Put yourself to this test. When the season is over and you are taking off the honey, notice how much more honey some colonies produce than others. Then get out your pencil and paper, and figure how much money you would have made if *all* colonies had made as much honey as the best. The results are frequently startling. Then remember that there is positively no one element that contributes to the production of these big yields as much as *good young queens*. After you have these results tabulated, consider whether or not it would pay you to rear your own queens and become an expert at it, or have some members of the family or firm take up this most important branch of beekeeping.

Our best authorities are agreed that there is not so much difference in the inherent honey-getting ability of the different colonies as there is in the *condition* of these colonies; that is, they produce large honey crops because the conditions within the hives are ideal. There were plenty of young bees and brood at exactly the right time. These colonies seemed to devote all of their energy to honey-getting. They did not loaf. They did not swarm. They just *worked*, and these conditions were brought about by the fact that these colonies had good young queens, and not because they had inherited any exceptional traits or were constitutionally superior. That there is a difference in the honey-getting ability of different colonies is not denied; but it is difficult, indeed, to be able to prove that the reason a colony made the largest surplus was due to natural ability rather than to the condition within the hive.

Therefore, it is no easy matter for the honey producer to pick out the best queens, since it may be the opportunity that the queen had, rather than her natural ability. How, then, are you to select your breeding queen? First, be careful to see that *conditions* are the same in all colonies, and that the queens are of the same age. Then select the queen that has the most desirable qualities, such as prolificness and vigor, and whose bees are gentle, of pure blood, good honey-getters, showing little inclination to swarm.

Years ago I endeavored to breed up a honey strain by simply using as a breeder the queen whose bees produced the largest yield. I found that the honey-getting quality was not in the least improved; but that the bees were getting cross and dark in color. Then I adopted the rule of selecting the largest and most prolific queen whose bees were gentle of good color. I found that better results were at once obtained. Being more prolific, this queen was able to keep the hive full of brood and the bees at the beginning of the honey flow, which is the secret of successful honey production. If this rule is followed and in addition all colonies are requeened from the best, in order to have them as nearly alike as possible in every respect, then we may select as our breeder the one that has the above qualifications and also the one that produces the biggest crop.

Some have reported that a medium-sized queen is as good as a larger one. That has not proved true in my experience, which has been that the larger the queen, the better. A queen that is extremely prolific has to be very large in order to contain the necessary number of eggs in process of formation to enable her to lay the four or five thousand eggs per day, which is the performance of a really good queen.

When the virgin emerges from the queen-cells she should be large, long and pointed. In three or four days, she will be much smaller, but extremely active and nervous. After mating she rapidly becomes larger until she is twice her former size. The abdomen becomes long and broad near the thorax, gradually tapering to a point. Short, blunt queens are inferior.

We must always bear in mind that, no matter how good our equipment, how well we pack for winter, how generous are the winter stores, and how abundant the nectar in the blossoms, our efforts will bring only failure if we do not have a good queen in the hive.

Chapter II. Conditions Under Which the Bees Rear Queens.

In order to rear queens successfully we must study the conditions in the hive under which bees rear their own queens. There are three of them known to beekeepers as the Emergency Impulse, Supersedure and Swarming.

In nature it sometimes happens that a colony suddenly loses its laying queen. Perhaps, as on very rare occasions, the queen, while laying, dies before the bees have time to supersede her in the regular way. The inmates of the hive at once realize they must meet this emergency, and immediately go to work to rear another queen. Fortunately, nature has made it possible for them to produce one from the larvae of eggs already in the hive. They, therefore, choose a number of worker larvae and begin to feed them lavishly with predigested food known as royal jelly. They usually fill the cell with this until the tiny larva is floated out to the mouth; then the bees build a queen cell over it, pointing it downward. This new cell is frequently over an inch long, and is made larger inside than that of the worker. The bees feed the larva until about five days after it is hatched from an egg, and then the cell is sealed over by them. The larva within spins a small thin cocoon, changes from a larva into a pupa, and in about eight days from the time the cell is sealed the virgin queen gnaws off the cap of the cell and crawls out. For a few hours she is a weak, frail creature, downy and delicate. However, she develops rapidly, and in from two to four hours, realizing she is a queen, she, just as many monarchs in the human family, becomes very jealous of any who may have ambitions to possess her throne. It is interesting to note the events which take place in the hive for the next few hours.

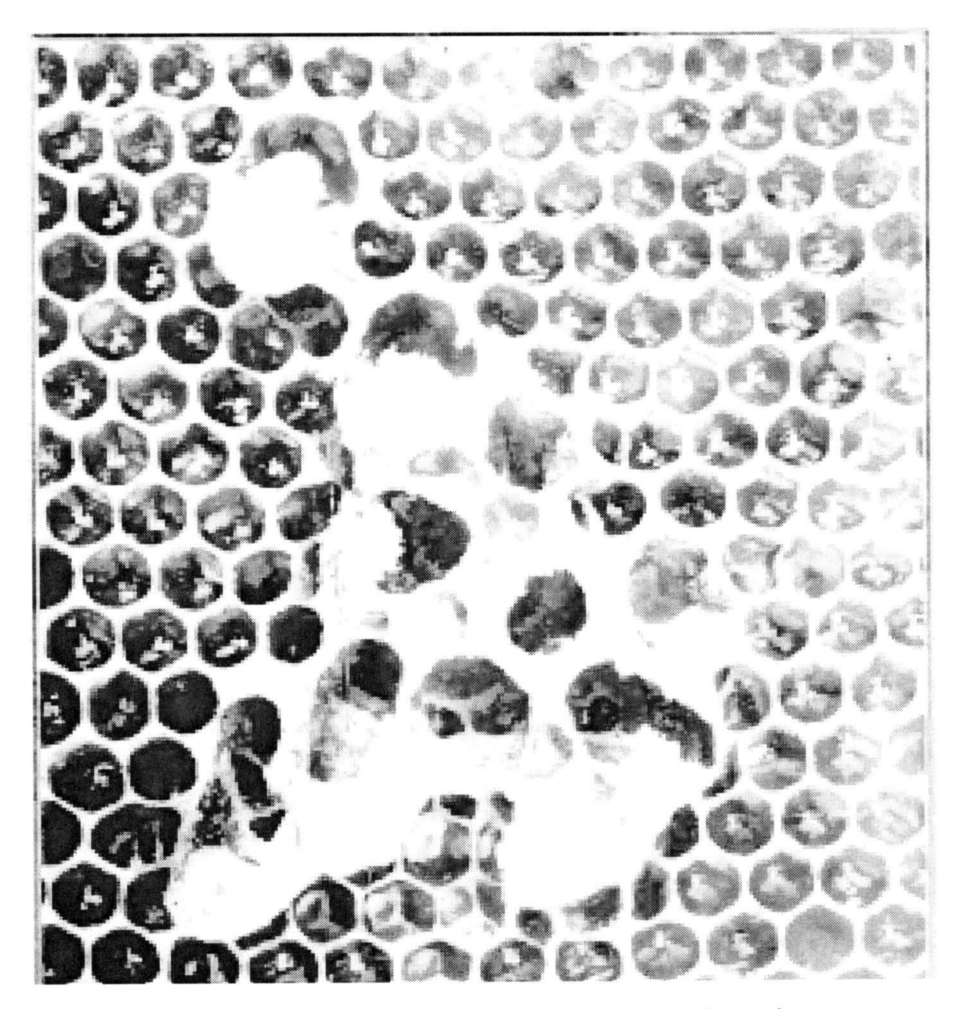

An opening is made in the side of each.

Having in mind the suppression of competition, the new queen roams over the combs. If there are any queen-cells from which the queens have not emerged, she supervises the destruction of them. The workers perform most of the labor under her directions, although she helps as best she can. She begins on the cells whose queens are most mature. She seems to reason these are the ones likely to give her the first trouble. An opening is made in the side of each, and, if the inmate is about ready to emerge, the queen backs down into the opening in the side of the cell and stings her helpless rival. The opening is then enlarged, and the dead queen is carried out by the bees. Other cells are visited and destroyed in turn. However, if there are queen-cells uncapped, these are left for a while, the newly emerged queen seeming to realize that she has plenty of time to handle their cases before they become any menace to her.

Now it frequently happens that, while this young queen is finding herself for the first two hours after emerging, other queens emerge, and several virgin queens will be in the hive at once. They seem to realize they are too young to do any satisfactory fighting, so by mutual consent they avoid each other's society and devote their time to supervising the destruction of queen-cells. However, as they grow from four to twelve hours old, they begin to seek out their rivals with the idea of doing battle. When they meet they clinch, and each tries to get a chance to sting the other. The fight does not last long, for soon one gets in the coveted position to give the fatal thrust of the sting in the thorax of her rival. The vanquished queen quivers a moment, and is dead.

Other "preliminary" fights are staged until only two queens are left. Then the "final" duel takes place, and the victorious queen reigns supreme.

In due time queen-cells in all stages of development are destroyed, and in six or seven or eight days the virgin queen flies out of the hive, meets the drone, and returns to become the mother of the colony, beginning her egg-laying within the next day or two.

A different order of events has been given by others, who state that the first thing a young queen does is to hunt up her rival and fight it out; but I have witnessed the occurrences many times as above described. Indeed, when occasionally grafted cells have been left too long in the hive, upon opening it I have found many queens safe and well, all busily engaged in tearing down cells. I have counted as many as fourteen superintending this work of destruction before any battle had begun. They have been given nuclei, thus saving them.

The most unsatisfactory manner in which bees rear queens is the Emergency method. The bees seem to feel their danger of extinction from having no queens. In their frenzy, a large number of cells are started. To make a bad matter worse, they take larvae that are too old, with an idea probably of rearing some sort of queen in the shortest possible time. (Transcriber's note, Jay changed his mind on this later and retracts this in Better Queens) We all know that in satisfactory queen-rearing, the younger the larva used, the better. By this method, the oldest larva chosen is the first to hatch, so the poorest queen in the batch is the one that heads the colony. However, as this is an emergency case, the bees seem to reason that, if this queen is not as good as she should be, they can take their time and rear a good one later on by the supersedure method.

Supersedure Method.

When a queen is beginning to fail from old age or some other infirmity, the bees seem to realize that she can not be with them much longer, so they take steps toward rearing for themselves a new mother. Queen-cells are started, sometimes only one, seldom more than four. In these shallow cups the queen lays eggs. As soon as hatched, the larvae are fed royal jelly, and as they receive the care and attention of the whole colony, good queens are, as a rule, the results. Sometimes the bees seem to wait until the old queen is so far gone that she lays several eggs in a queen-cell which results in the larvae not having sufficient food since they have to share it with their "cell mates." Owing to their being crowded in the cell, such queens are sometimes slightly misshapen. Usually, however, all but one of the larvae are removed before the cell is sealed. Under the supersedure method, however poor queens are rare and, as a rule, the best of the queens are reared. Usually the old queen disappears as soon as the virgin emerges from the cell; but sometimes, mother and daughter live peaceably together, both laying and usually found on the same comb.

Queens Reared Under the Swarming Impulse.

When a colony is preparing to swarm they start a large number of queen-cells in which the queen lays eggs. When the first cell is capped, if the weather is favorable, the swarm usually comes out. As swarming occurs when the colony is at its height of brood-rearing, the larvae are well supplied with royal jelly, so that the finest queens are reared. In rearing queens by any method, we can learn a great deal by carefully studying the conditions of the bees while building cells preparatory to swarming, for we wish to duplicate the performance.

Under the Emergency method, the bees build a large number of cells, but they do not give them the proper attention and skimp the larvae for food. Under the Supersedure method, they give the larvae plenty of food, but usually do not build more than three or four cells. Under the Swarming Impulse, they not only build large numbers of cells but supply the larvae in them lavishly with food. What is the reason for this? Is it because they have the "swarming fever" that induces them to do such good work? I believe not. My observations lead me to believe it is the *condition* of the colony and, in support of this theory, I have found that as many and as good cells may be built by a colony when not preparing to swarm as by one that is, provided the conditions are the same in all other respects.

What are these conditions? First a honey flow is on or just coming on, for bees seldom swarm at any other time. Second, they are strong in bees, especially young nurse bees. Third, the hive is crowded with brood in all stages; and fourth, the weather is reasonably warm. I believe these conditions enable the bees to rear not only a large number of queens but those of the highest quality. Understand, it is the *condition not the swarming fever*. As evidence to substantiate this statement, the following fact, which I have observed many times, is given. While having cells finished above an excluder, sometimes the bees take it into their heads to swarm, and as bars of cells are capped the swarm issues. Since the wings of the queen are clipped the bees return, and the queen is helped back into the hive. Removing the bar of cells frequently discourages swarming but sometimes they persist coming out every day or every other day for a week or more as the spirit moves them. I have never been able to see that, while they had this swarming fever, they gave the cells any better attention than before or after swarming. This fact satisfies me that it is the condition of the colony and the honey flow or the feeding that give good results in cell-building.

Under the Grafting method, we endeavor to get all colonies connected with queen-rearing in the condition above described. If we do, we can rear queens every bit as good as those reared under the swarming impulse; (Transcriber's note: Jay later changed his mind on this.) if we do not, inferior queens will result. By examining the cells one can easily tell which of the three methods the bees used in their construction. In the Emergency method, the queen is reared from a larva that has hatched in a worker-cell, so by looking into the bottom of the queen-cell, the worker cell may be seen. In the Supersedure method as well as the Swarming method, the cells are the same. The queen lays eggs in both; but during the swarming, many more cells are built than under the superseding impulse.

Queen-rearing apiary of the author.

Chapter III. Queen Rearing for the Small Beekeeper

There are several methods that may be employed where one wishes to rear but a few queens. Cells, saved from a colony that has just swarmed, may be placed in colonies to be requeened, whose queens have been removed. This is much better than to allow colonies to run along with inferior queens; but, by this method, little progress can be made in improving the stock since when you wish to requeen, your best colony may not be swarming. Consequently, you would have to use cells from an inferior colony. It has frequently been noted that the inferior strains of bees swarm the most. Blacks and hybrids are much more inclined to swarm than Italians.

In requeening by the swarming method, a piece of comb one inch in diameter should be cut out around the cell, using a good sharp knife, and being careful not to injure the cell. A hole of corresponding size should be cut in the comb of the colony to be requeened and the piece containing the cell fitted into it. Where but one cell is on the comb, the entire comb may be placed in the colony to be requeened. If this colony is of medium strength or strong, it makes no difference just where the cell is placed for there will be sufficient bees to give it proper incubation. The bees may be left on this comb or brushed off, but never should be shaken off since the undeveloped queen is almost sure to be injured. In giving a cell to a weak colony or a nucleus, it is important to place it near the center next to the brood. Frequently cells built on the bottom edge of a comb when given to a weak colony do not mature.

A second and very simple method of requeening is simply to remove the queen from a colony, and the bees will construct a number of cells by the Emergency method. Such queens are not, as a rule, as good as those reared under the Swarming or Supersedure Impulse. If care is taken to save only the largest and best cells, however, very good queens can be reared in this way. The principal point to commend in both of the above methods is their simplicity. If one has never reared queens, these will prove very interesting and are a step toward better ones.

The third system requires a little more skill, but will produce cells as good as the best if care is taken to have all conditions right. Go to the colony containing your breeding queen and insert an empty comb into the center of the brood-nest. Leave this there for two or three days or until the queen has laid a large number of eggs in the cells. Remove it, however before the eggs begin to hatch since our object is to get the bees to use very small larvae from which to rear queens.

Next, go to a strong colony and take the queen and all combs containing eggs or brood, but leave with the bees several combs of honey and pollen and give them the frame of eggs from your breeding queen. If it is desired to save this queen, she is given a frame of brood and adhering bees and put into a hive to start a new colony. Fill out the vacant space with combs containing some honey, if possible. If you have no extra drawn combs on hand it is better to take a few from other colonies and in their place give full sheets of foundation, for they will do better work at drawing foundation than would this new colony which is not strong enough for that purpose. The remainder of the brood is used to strengthen weaker colonies or to make strong colonies even stronger for the honey flow as occasion seems to demand.

Having now disposed of the queen and brood, let us go back to our queenless colony. Realizing their queenlessness, the bees will start cells as soon as the eggs begin to hatch. Very frequently by enlarging the worker-cell, they make it over into a queen cell even before the egg hatches. In this manner the newly hatched larvae receive abundance of royal jelly from the very start, which is necessary for the best results. This method has the advantage over the others just described since the bees can not use larvae that are too old for good results. However, it should only be used when there is a honey flow. In about six days after the cells are capped, they should be cut out with a sharp knife and given to colonies to be requeened which have been made queenless. When there are larvae of the proper age at the bottom of the comb, the bees prefer to build cells there, sometimes building a compact row of cells half way across the comb. In such cases some of the cells will have to be destroyed when being cut apart. In giving this comb of eggs to the colony, if there are no eggs at the bottom of the comb, it is well to cut away the comb so that the eggs will be at the edge. This is not necessary, however, for the bees will start plenty of cells if the comb is left intact. As the operation of forming nuclei to receive them, when that is desired, is the same as given under the Grafting method, it will not be described here.

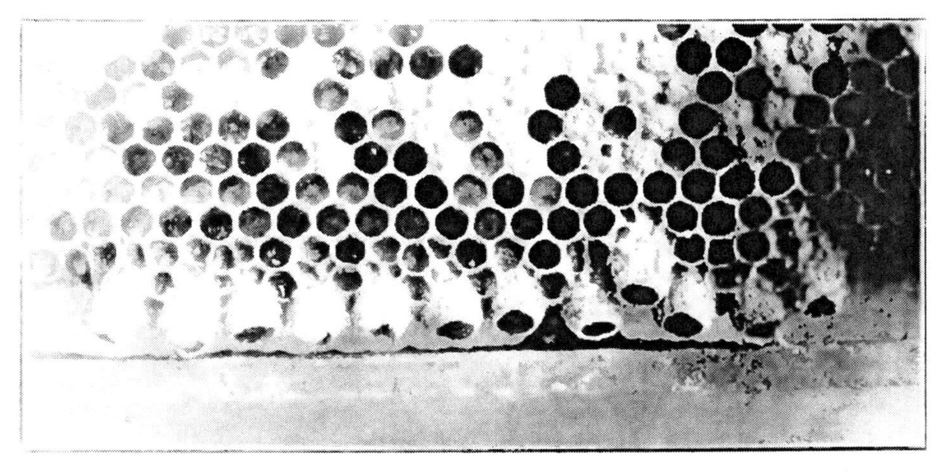

A compact row of cells half way across the comb.

The Grafting Method.

If one keeps as many as fifty colonies or expects to do so in the future it will pay to learn the grafting method. This requires much more skill and practice than the ones above mentioned; but it has so many advantages over all the rest that it is used by nearly all queen-breeders and extensive honey producers who rear their own queens.

This method is more economical, for it is not necessary to have any colony queenless at any time. You have control over the situation and can rear queens in any quantity desired. It is exact, since you know within a very few hours when any cell will hatch. The artificial queen cups are much easier to handle, for with them it is not necessary to cut up and mutilate good worker combs. Larvae can be taken from your best breeding queen and the stock improved thereby. Last, but not least, the very best queens can be reared, if conditions are kept right. To rear a few queens during a honey flow is a simple matter; but to keep up a steady production throughout the season under variable weather and honey flow *is not* a simple matter. However, with experience and patience it can be done.

Root's Basswood Apiary.

I thoroughly believe that many beekeepers who have a thousand colonies or more and who do not rear their own queens could increase their honey yield fifty per cent by having a good queen-rearing outfit and being able to use it properly. Moreover, in localities where European foul brood is rampant the honey crop might be doubled or trebled, since there is nothing that eliminates this disease like strong colonies of Italian bees headed by young, vigorous queens.

Headwork.

Some of the most important work that can possibly be done in the winter months is reading bee books and journals, of which we have a goodly supply of the highest standard. Secure all the books you can and take all of the journals. If you do this and carefully study them, it will be the best investment you can make. Read, study and plan in the winter.

We should remember that successful business men work with their *heads.* They can hire hand work at a low figure; but headwork is always at a premium. A great deal of headwork is required of the successful beekeeper, and much of this work can be done in winter. During the honey flow we are too busy working with our hands to do much headwork. J.S. Knox, the efficiency expert, says that a man is worth $2.50 per day from his chin downward. If he earns more than this, it must come from above the chin. Consequently, he divides men into two classes, "Chin Uppers" and "Chin Downers." If we are successful we must be "chin uppers." For the beekeeper the best time to do his "chin upper" work is in the winter sitting before a comfortable fire, reading, thinking, studying, planning.

Moreover, as there is a great deal of work to be done with the bees during the queen-rearing season, one should plan to do all the work possible in the winter. Nailing up hives and nuclei, painting them, putting in foundation, dipping cells and similar work should be all gotten out of the way before spring comes.

Chapter V. Dipping Cells.

Since I know more about the way *I* rear queens than I do about the way any one else does it. I wish to take the reader with me through the season, while I attempt to show in detail how I rear queens. Possibly, you have methods of your own that you prefer. I do not claim to have a monopoly on all the good things in queen-rearing, but will be content if you find some little feature which I use that you consider worthy of adoptions, and which may be of help to you.

Let us start by dipping cells as this can be done in the winter. Wax is saved from the year previous. For this a solar wax extractor is an important item. During the summer months, many small pieces of comb are found that can be thrown into it. This makes the finest cell building wax. In the nuclei, bits of comb are built and when introducing queens, where a frame is taken out, the bees will construct more or less comb. All these can go into the wax extractor. From the wax extractor, the wax is placed in small molds, for use in dipping queen-cells. I have enough cell bars to last the season, so we always dip sufficient each winter to supply us through the entire summer.

Our cell-dipping outfit contains twenty cell-forming sticks, which work through holes made in two pieces of heavy tin. Metal is much better than wood since the latter swells when wet and the forming sticks do not work freely through the holes. These pieces of tin are fourteen or fifteen inches long, fastened one and one-quarter inches apart to small blocks of wood, which are to serve as handles when dipping the bars into the trays. Each piece of metal is pierced with twenty holes, one-fourth inch apart, and seven-sixteenths inch in diameter. The holes are exactly opposite each other on the two bars, in order that the cell forming sticks may slip up and down through them easily.

A solar wax extractor is an important item.

Two trays are used, one five by sixteen inches, the other two and one-half by fifteen inches. Water is placed in the larger forming a double boiler; while wax is placed in the inner tray and the whole set over the heat. The wax should be kept at the lowest temperature at which it will remain liquid. If it becomes too cool the cells will be lumpy; if too hot, they do not slip from the sticks. If one is not experienced, it is well, when the wax apparently reaches the proper temperature for successful dipping, to try dipping one stick, and, if the wax proves of satisfactory temperature, proceed to work.

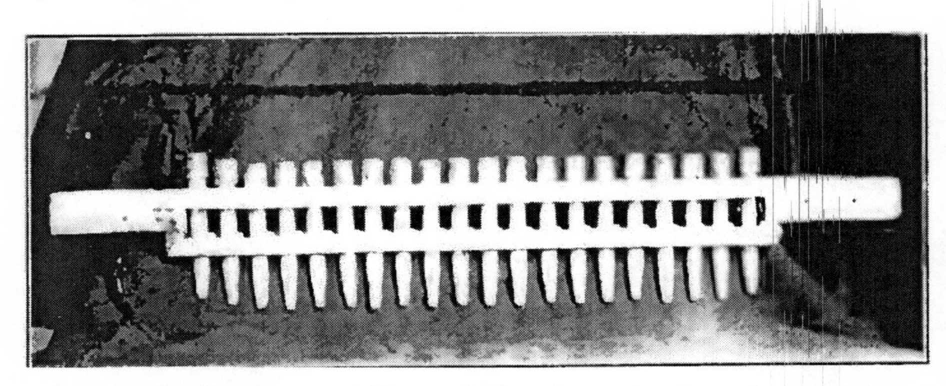

Our cell-dipping outfit contains twenty forming sticks.

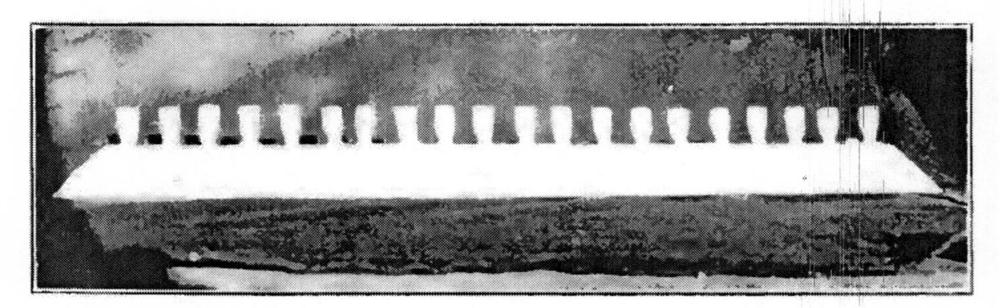

Cells of the proper size and shape.

First, dip the ends of the forming sticks in cold water, then dip into the melted wax; again dip in the water and back into the wax for about four dippings, care being taken to have a firm thick base, with a thin even edge. By dipping the sticks in the wax and holding the bar up until a drop forms on the base of the cell, a thick base is procured. A thick base is necessary, for in trimming off the cells with a knife the cells would be injured if too short. When completed, the cells should be about five-sixteenths of an inch across the mouth and one-half inch deep inside measurements.

Many beekeepers make a mistake in believing that the most important feature for successful cell acceptance is the grafting of the larvae into the cells cups; but a far more important feature is that of making cells of the proper shape and size. The ideal cell would be as the bees build them, large inside, with a small mouth; but it is not possible, or at least practical for the beekeeper to make cells of this shape. Upon several occasions, I have given cells that had been accepted and slightly built out in the swarm box to a colony for finishing, when by accident it contained a virgin queen. Of course, the larvae and jelly were both quickly cleaned out. I have given one bar of such cells to a swarm box and two bars of our dipped cells. The bees seemed to concentrate all their efforts on the cells already worked on by the bees and neglected my dipped cells. The bees prefer to make the mouth of the cell just large enough for a worker bee to crawl into, and it is frequently noticed that sometimes in the workers haste to back out of a queen-cell when smoke is blown into the hive, it is caught and has to do considerable scrambling and kicking before it can get out. I find the best cell for practical purposes is one whose size is between that of the inside of a natural queen-cell at its largest place and the mouth of the cell, this being five-sixteenths of an inch as given above. In our early experience, many of us, enthusiastic in rearing larger queens, sought to accomplish this by making larger cells; but being large at the mouth, the bees were loath to accept them, and it took considerable work on their part to build them over to the size they should be. When the bees get to work on the cells they mold them into the shape they want, regardless of the size and shape the beekeeper has made them. The smaller cells will give better acceptance than the larger ones; but do not for a moment imagine this cramps the larva and produces an inferior queen, for the bees enlarge the cell to suit their own fancy. For experimental purposes I have dipped queen-cells the size of a worker-cell, and excellent results were obtained. Cells larger than five-sixteenths of an inch are not accepted so readily as those of this size or smaller.

And the cell cups painted at the base.

Nothing but pure beeswax of good quality should be used. Upon one occasion, when everything was going finely, cells accepted and built out nicely, the bees in the swarm boxes began to balk until accepted less than twenty-five per cent of those given. I had all conditions right, as I supposed, the same as before-plenty of young bees, well fed. At length I noticed the wax of which we made the cells was not so white as some we had been using. I made up a new batch of cells from clear white wax, and as if by magic, all cells were again accepted and everything went on splendidly as before. Instead of heating the wax in a double boiler as we do now, this wax had been set directly over the flame and had become slightly scorched and darkened, so the bees would have none of it.

After the cells have remained in water long enough to become slightly hardened, they are loosened by giving each a slight twist, but allowed to remain on the sticks. They are then placed on the cell bar, the frame being supported on blocks. A small round paint brush is dipped in hot wax, and the cell cups painted at the base where they come in contact with the cell bar. A kettle should be kept at hand for melting additional wax to add to that in the inner tray, in order that sufficient wax may be had to make the cells the necessary one-half inch in depth. If the wax in ether becomes dark-colored or impure it should be discarded, and an entire batch of new clear wax placed in the tray. However, the darker wax may be used to paint the bases of the cells to cause them to adhere to the bar.

When the wax has become thoroughly cool, the frame is lifted off and all of the forming sticks come out of the cells easily. If properly done, the cells will remain on the bars even if subjected to considerable rough usage. When the cell bars are all finished they should be wrapped carefully in paper to be kept free from dust, since the bees will not accept dirty or dusty cells. If you have on hand the cardboard cartons in which foundation is shipped they make ideal containers for the cell bars.

Suggestions in Making Cell Cups.

Of course it is not advisable for the beginner to have a dipping outfit made as previously described. After mastering the grafting method, he may enlarge upon his equipment as he wishes.

The beginner can either dip his cells one at a time and mount them or he can purchase ready-pressed cells from dealers in bee supplies. Either one will give perfect results. These cells may be mounted on bars as needed, thus eliminating the necessity of purchasing a large number of bars. The base of these cells may be dipped in hot wax and stuck on to the bar when needed. To avoid the necessity of getting the swarm box, he can also use the queenless and broodless method described in Chapter XIII. However, I believe it pays to use the swarm box, for one can, as a rule, get better results. In this way it is possible to experiment until one gets his hand in without putting much money into equipment, and as he progresses can add to the equipment to fit his requirements.

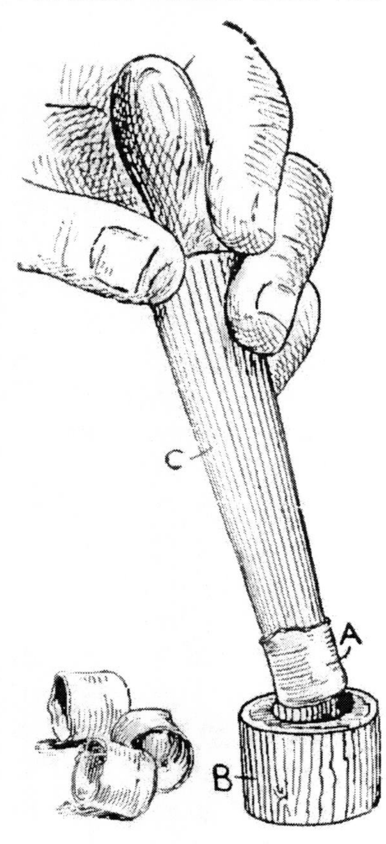

Pressed cell cup.

If one has difficulty in making his cells, one at a time or collectively, he can use to advantage the ready-made pressed cells sold by all dealers. Where only a small number are required the beginner will probably do better to buy what few he uses. The making of dipped cells is a nice art, and unless they are made just right, the bees will reject them.

Chapter VI. Royal Jelly.

Every thing in readiness, we await the coming of spring with a great deal of enthusiasm and no little impatience. Sometimes it seems spring weather will never come; but it does no good to worry and if you look backwards, you remember that spring has never yet failed to come.

Is there anything more interesting than to watch the bees bring in their first loads of pollen? If the beekeeper has done his duty toward them the season before, there will be no need of tinkering with them until later in the spring. The soft maples blossom and go; then come the pear and apple blossoms, and soon a few heads of white clover can be seen. It is now time to get busy at queen-rearing. Some seasons the weather permits grafting soon after the first blossom; but it does not pay to be in too great hurry to rear queens before the real queen-rearing season arrives. I know of no definite rule concerning the time for it. Each person will have to find out by experimenting until he knows his location well enough to be reasonably sure when to begin.

Many times in the Mid-West, the bees are strong and the weather conditions ideal for cell-building during apple blossoms; but later the weather turns cold, so that virgin queens can not get out to mate. As nearly as I can come to it, when the hives are getting nicely filled with brood, when plenty of pollen is coming in and the bees are gathering a little nectar, then it is time to begin grafting.

Before grafting, a supply of royal jelly is necessary. Some very successful queen-breeders report they get satisfactory results without its use; but I have never been able to procure as large acceptance or as good strong queens without it. *(Transcriber's note: Jay Smith changed his mind on this in* Better Queens.*)* J.W. George of El Centro,

California, gave to the beekeeping fraternity a valuable little kink when he explained that royal jelly can be bottled and kept in perfect condition from one season to another. I have practiced this to advantage, and find one of the great difficulties of queen-rearing is thereby removed.

If you have no royal jelly on hand, a colony may be made queenless until they build queen-cells, when you can get the jelly from them. After the first grafting, some of the jelly in a few cells you have produced may be used; but, in this way you continually destroy good queen cells.

As a container for royal jelly, I use a small porcelain jar with a screw cap. A piece of waxed cardboard in the cover makes it air-tight. Let me offer a suggestion as to where you can get one of these jars. Make a raid on your wife's manicuring outfit, and, if luck is with you, you will find one of these jars. To be sure that luck will be with you, better do it when she is out. This jar usually has some pink dope in it. Take this out, put it into a tin can, present it to your wife with your compliments and make off with the jar. Thoroughly sterilize this jar by boiling, for the bees seem to object to the funny smell that comes with it. If your wife does not have this, or if you do not have a wife, you can go to the drug store and find just the size and style that suit you. The dope looks as though it might be of use if you put it into the grease cups of your flivver, but I do not want to suggest too many dangerous experiments for you to try all at once. For a jelly spoon, I prefer to make one out of the bone handle of a toothbrush, which also may be found in the manicuring outfit. Break off the brush and whittle down the small end until it fits nicely into a worker-cell. This jelly spoon and the jelly jar are to be carried in the pocket of your trousers or dress, whichever you wear. While working with your bees during the season you will be running across colonies that have royal jelly to spare. Whenever a swarm issues, just take out the jar and spoon and get the royal jelly. I have found that I come across enough in my regular work so that I never have to make any special hunt for jelly. It is well to have two of these jars; keep one in your pocket and the other in the grafting room.

And get the royal jelly.

Chapter VII. The Swarm Box.

There are several methods used for getting cells accepted and started by the bees; but when all things are considered, I believe the swarm box has more desirable features than any other. For best results the swarm box must be kept in even temperature. It gives perfect results during cool weather, for, by placing it in the cellar, cave or basement, the outside temperature does not affect the bees. During hot weather it gives equally good results for the same reason. By using the swarm box it is not necessary to have any colony queenless at any time.

The swarm box.

The box is made eight inches wide, inside measurement, and should be the right length and depth to take the sized brood-frame to be used, allowing about an inch space below the frames. The bottom is covered with wire-screen cloth. Four legs one inch long are fastened to the bottom in order to provide plenty of ventilation. At each end of the box two strips of wood are nailed to support the cell bars. These should be placed so that the cell bars will be a little lower than the top of the brood frames. A cover is made with cleats running entirely around, forming a telescope cover one inch deep. In the center of this cover is made an oblong opening large enough so that three cell bars may be passed through it with ease. The object of this cover, which is little more than a rim around the edge, is to prevent the bees from boiling out over the sides when the bars are being put into the swarm box. A second cover slides on top of the telescope cover, thus closing the opening in it. The end-cleats on the telescope cover extend a little above to keep the upper cover in place. Two Heavy wire handles swing up over the ends so that when the swarm box is being carried, these covers cannot fall off. A round opening is made in the top of the upper cover by the use of an expansive bit just the size to take the cap of a mason jar. When feeding the bees, a Mason jar with a perforated cap is used in this opening which is also utilized for putting the bees into the swarm box by inserting a tin funnel and shaking the bees from combs into the box. The box is given two coats of paint to prevent it's warping so it will thus remain bee-tight. When made in such a way as to have the cell bars inside, the bees may cluster all around them and thus keep the temperature uniform, which gives much better results than the old style where the bars were placed in slots in the cover.

The old style swarm box is somewhat easier to make and for experimental purposes might be preferred by some. It is similar to the one just described except the cover, which is merely a thin board with an opening cut in the top large enough to hold the three cell bars. In using this it is well to place a cushion over the bars to keep out light and retain the heat. In this style it is necessary to make the filling hole at the end of the cover to avoid the opening made for the top bars

These covers cannot fall off.

Chapter VIII. Getting the Bees in Condition for Cell Building.

Let us remember that for the best results in cell-building we must have plenty of young bees which are being lavishly fed either from a honey flow or from receiving sugar syrup. This condition is necessary where even a few cells are being built by the colony. Now, as we wish them to build a large number of cells, the colony must be *exceedingly strong.* As we are starting early in the spring while only a little nectar is coming in, it will be necessary to feed sugar syrup in order to get the best results.

The method of feeding that I have found very satisfactory is to take a two quart mason jar and punch in the cap eight nail-holes 1-16 inch. Fill it with syrup, equal parts of granulated sugar and water. Many recommend a weaker syrup, but with me the heavier syrup gives the best results. There is no loss as the bees store in the combs all that they do not need. A bee-escape board is used for a hive cover, and the Mason jar is inverted over this hole. An empty hive body is set on, and a regular hive cover is placed on top of all. By this method of feeding there is no robbing, and the bees take the syrup night and day even if the weather is quite cold, which they will not do when this style of feeder is used at the entrance. However, our regular bottom-board feeder, as described later, has so many advantages over any other that I am now using it for all purposes.

As the swarm-box colonies as well as the finishing colonies are very strong, it is an advantage to slide the hive forward on the bottom-board, thereby affording better ventilation. This also provides an opening into which the syrup is poured. Many advocate giving them a very little thin syrup from an entrance feeder, using about a pint a day. This will answer very will if some nectar is coming in; but, when this is not the case, better results can be obtained by giving syrup in abundance. Two quarts of syrup, equal parts sugar and water, per day will give excellent results. Enough must be given to cause the bees to build white comb, and this enables them to draw out the cells to perfection as well as to secrete an abundance of royal jelly. My experience has been that the bees do every bit as well when the feed is given them all at once every night as they do when they take it through three or four holes from the Mason jar. It must be borne in mind that there is no waste in giving them more than they use at the time, for all surplus is stored in the combs. The colony that is to furnish bees for the swarm box must be very strong in bees. If it is not in this condition, it must be built up by giving it frames of emerging brood from other colonies. If a hive smaller than the ten-frame jumbo is used it should have a double brood-chamber, and both stories should be full of bees and brood. The hive should contain ten or twelve pounds of bees. Rearing good, vigorous queens without strong colonies and plenty of feed is an impossibility.

By this method of feeding there will be no robbing.

The colony that is to furnish bees for the swarm box must be fed at least three days before the swarm box is filled. It will do little good to feed them just before they go into the box. I do not know why this is true, but it seems to take a few days for the bees to assimilate the food and make it over into royal jelly. So I use the term, "fat bees." You must fatten the bees before they can do good work at feeding larvae. Poor, hungry bees will not accept cells. If there is a pretty good honey flow on, no feeding will be required.

This also provides an opening into which the syrup is poured.

Chapter IX. Cell Finishing Colonies.

Since we have the swarm-box colony in fine condition with abundance of bees and supplied with food, we will prepare the finishing colonies. As we expect to start three bars of cells in the swarm box, it will be necessary to prepare three colonies to finish them, for one bar of twenty cells is enough for even the strongest colony. However, if the finishing colonies are sufficiently populous, they will do exactly as good work at finishing twenty as can be done by a colony preparing to swarm, in building cells in their own natural way. The method of preparing the finishing colonies is similar to preparing the swarm box colony. They must also be kept running over with bees. At the beginning of the season when the colonies have not had time to build up to maximum strength, a large amount of brood is required to put them in condition to do the best work. All empty combs in the brood-nest should be removed and replaced with brood from other colonies.

They must make a two-story colony. The second story must be added with a queen-excluder between the two hive bodies. If the colony is of sufficient strength to care for nine frames of brood, the frames are placed in the upper story after shaking off all the bees. If the finishing colony is not strong enough to take care of the extra frames of brood, it is best to give it the nine frames of brood with adhering bees. There is some danger of these strange bees killing the queen below, and in order to prevent this, place a newspaper between the two bodies on top of the queen-excluder and let them unite the same as when two colonies are united. In this case you have the advantage of both brood and bees. In two weeks most of the brood will have emerged, and the combs will be filled with sugar syrup or honey. Remove these and put in some more brood. Do not wait until you notice that the cells are not being finished as they should be, for, if you do, a lot of inferior queens will result.

Running over with bees.

These combs of honey with a little capped brood are excellent for giving to colonies that are short of stores, or they can be given to nuclei. Always keep unsealed brood on each side of the frames containing cells, in order to draw nurse bees to them. No matter how strong the finishing colony may be, it will do poor work at cell-finishing unless there is unsealed brood in the upper story.

We must bear in mind that, when feeding is necessary, it must be done several days before the bees are to build cells-two days at the shortest and three days are better, the same as with the colony that is to furnish bees for the swarm box. If fed three days before going into the swarm box, they will be in splendid condition to feed the larvae and to draw out the cells in the proper shape.

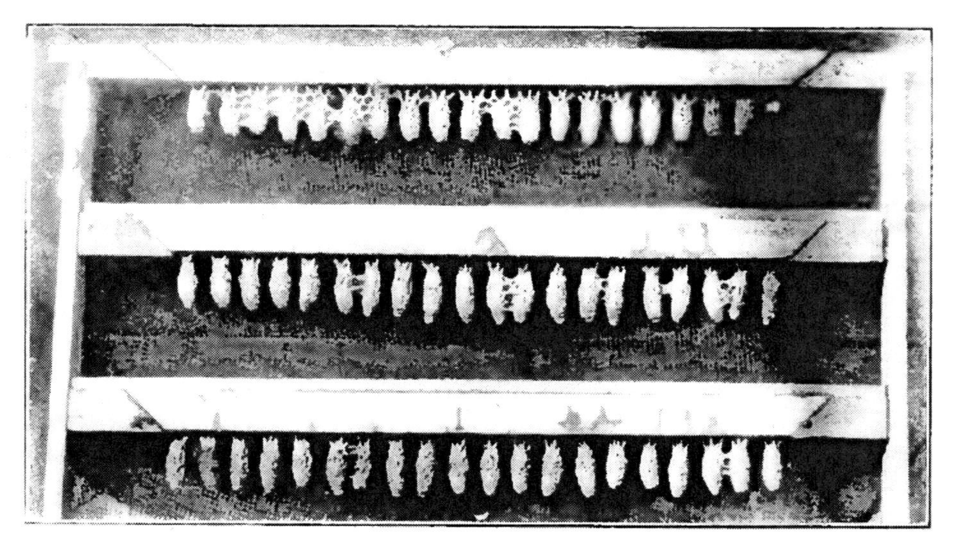

Showing September cells.

The very best of cells may be built any time of the year after brood rearing is well under way in the spring until it stops in the fall. In order to have good cells built out of season, it is necessary to put the colony that is to supply the bees for the swarm box and the finishing colony in the proper condition. We should remember it is the *condition* of these colonies that brings results, and not the time of year, the honey flow or swarming fever. Therefore, if we build up the colonies with brood taken from other colonies and feed these built-up colonies, we have as good conditions for cell-building as we have with a strong colony during a honey flow. It is evident that to build up colonies to cell-building conditions in early spring or late fall is expensive, as it will rob a number of colonies of their brood; but, if queens are needed at such times, it can be done with profit. The illustration shows some bars of cells built in late September after the flowers have all gone. But few of the virgins that emerged from these cells ever became laying queens for the weather turned cold and they could not venture out on their honeymoon.

One should bear in mind that it is the *best* queens that make the records at honey getting, so it pays big dividends to be extravagant with brood and feed. If all cells have dried-down royal jelly in them after the queen emerges, you have done all that can be done in the way of providing bees and feed to the cell-building colony. If some cells have no jelly in them, you are not rearing the best of queens. True, some may be first class for they may have had enough after consuming it all, but there are sure to be some that do not have enough and dwarf queens will result.

How many cells can a colony finish? During the summer of 1923 some experiments were made at our yard to determine the number of cells a colony should finish. In stating the number in this book, we wish to stand on firm ground and not advocate anything that might bring poor results to the beginner. If our experiments prove conclusive, we shall give them to the public later; but at this writing I believe that, when the colony that is to start or finish the cells is in perfect condition, it will build a large number of cells, and when not in good condition, it will not do good work on one cell. This is somewhat similar to the perplexing problems of "over stocking." When the honey plants are in good conditions, overstocking is almost impossible; but, when the plants are not in proper condition, a single colony can not make a surplus.

Root's grapevine apiary.

Chapter X. Filling the Swarm Box.

Two combs containing some honey and pollen are placed in the swarm box. These should be old combs and not too heavy, for, in the handling they are to receive, they will be liable to break down if new or if they contain much honey. These combs are placed one at each side of the box and are held in position by the two blocks that are to support the cell bars. If one has never used a swarm box, it is well to place it on scales for a few times until able to judge accurately the weight of the bees the box contains. A funnel such as is employed in the filling of pound packages, is used for putting bees into the box.

While good results can be had with no pollen in the swarm box, better results are obtained by having plenty of it in the two combs that are used. It is surprising to note the amount that the bees will consume while confined in the swarm box. If the two combs have an abundance, it will usually be eaten after the combs have been used three times. Before filling the box with bees, examine the combs, and, if they do not contain sufficient pollen, they should be removed and other frames containing plenty put in their place.

Set the swarm box in front of the colony from which the bees are to be taken, put the funnel into the hole and all is ready for the bees. It is quite desirable, upon all occasions when removing frames from the hive, to see that they are put back in the same position as found. If not, queen-cells are apt to be started, and when the virgin emerges, she will kill the laying queen. This subject will be discussed more fully under "Introducing Queens." A good method is to take out the frame nearest you and set it several feed away from the entrance. Then examine the next frame to find the queen. If she is not on that frame, set it back, lift out the next and then the next until the queen is found. Pick her up by the wings and put her on the frame you first took out. Set all frames back in their regular place except the one that has the queen.

The reason we set the first frame with the queen some distance away is to prevent the bees and the queen from crawling back into the hive, thus getting the queen into the swarm box.

Give it a quick shake downward and then upward.

Nothing I can call to mind creates such a strong desire to kick one's self as to get the queen into the swarm box. I know from experience. While you are working with the bees, they begin to fan and the first thing you know all the bees, including the queen, begin a grand march for the entrance or go over the top and into the hive. You do not know that the queen is there, and your grafting comes to naught. Instead of accepted cells you find them mostly torn out by the roots and made over into some fine worker comb containing eggs. In order to avoid this calamity, set the frame with the queen so far away that the bees will not heed the call of their companions.

We are now ready to put in the bees. Take out the first frame covered with bees, put one end of it down into the funnel, take hold of the other with both hands and give it a quick shake downward, then upward. Two little shakes, in which the comb is not moved more than two or three inches, will dislodge all of the bees except those that have their heads down in the cells. Set this frame back into the hive and do the same with the others until the desired number of bees are in the box. From five to seven pounds is the right amount. Try to get six pounds as nearly as possible. With a little practice you will not vary much more than a pound either way. There should still remain in the hive a sufficient number of bees to care for the brood. When the box has the required amount of bees in it, remove the funnel, place the cap of a mason jar in the hole, replace the comb containing the queen, close the hive and carry the box to the basement.

Caution. When a heavy honey flow is on, take care that the bees are not daubed with honey when they are shaken into the swarm box, for if they are they will suffocate and both bees and cells will be lost. True, if a little honey is smeared on them it does no harm; but too much is disastrous. When a heavy honey flow is on, shake the comb lightly so no nectar is displaced, and, if sufficient bees are not obtained in this manner, the bees not shaken off may be brushed off with a bee-brush. When more than five pounds of bees are put in the swarm box, it is advisable to set the box on two by four scantlings to afford more abundant ventilation.

Are contented and satisfied as though in their own hive.

The Dungeon.

In one corner of the basement I have what I call the "bee dungeon." This is a room made by stacking up extracted supers and hive bodies to the ceiling to make it dark. The opening that serves as a door is made in break-joint style so that no light can get in. It is wide enough so that a person can walk in carrying a swarm box in each hand, which is another advantage over a swinging door for plenty of fresh air can enter. Back in the dungeon the bees remain quiet as though it were night, away from noise, light and strong air currents, and are as contented and satisfied as though they were in their own hive. If no basement is available, any room in the honey-house where it is not too hot or cold will do; but it will pay to make a basement. It is the ideal place. I usually fill the box at one o'clock in the afternoon and leave the bees confined there in the basement until three o'clock. I find that two hours of confinement is all that is necessary, for as the bees are queenless, broodless and on strange combs, they realize their queenlessness to the fullest extent in that length of time.

Why the Bees Accept the Cells.

Now, while the bees are contentedly clustering to the lid of the swarm box, licking the honey off any luckless individual that was daubed up when they were shaken from the combs, let us consider the condition brought about with the bees that causes them to do good work at cell-accepting. For some days previous, the young nurse bees have been feeding great hoards of larval food which is the same as the food we call royal jelly. We have suddenly taken them away from these larvae, so they continue to secrete the royal jelly but have no larvae to feed. They also realize their queenlessness. They are crying for a queen; they have the food with which to raise many queens, but they have no larvae with which to do it. It is our privilege to accommodate them in this respect, so now we will proceed.

Chapter XI. Grafting the Cell Cups.

The best place to do the grafting is in the honey-house or the room of a dwelling where there is plenty of light coming through a south window. A room is better than out of doors for several reasons. It is cool, and the larvae may be kept away from strong light, heat and drying winds. It is more comfortable for the operator, and he is away from robber bees. The grafting outfit is quite simple-a grafting needle that can be bought from dealers in bee supplies, a jelly spoon made out of a toothbrush handle, a little jar of royal jelly and a small individual salt dish in which to mix the jelly. With the jelly spoon, place some of the royal jelly in the salt dish and dilute it with pure water. It should be as nearly as possible like the thin larval food seen in the bottom of the worker-cell soon after the egg has hatched. When this is done, go to the hive containing your best breeding queen and take out a frame with as many young larvae of proper age as possible.

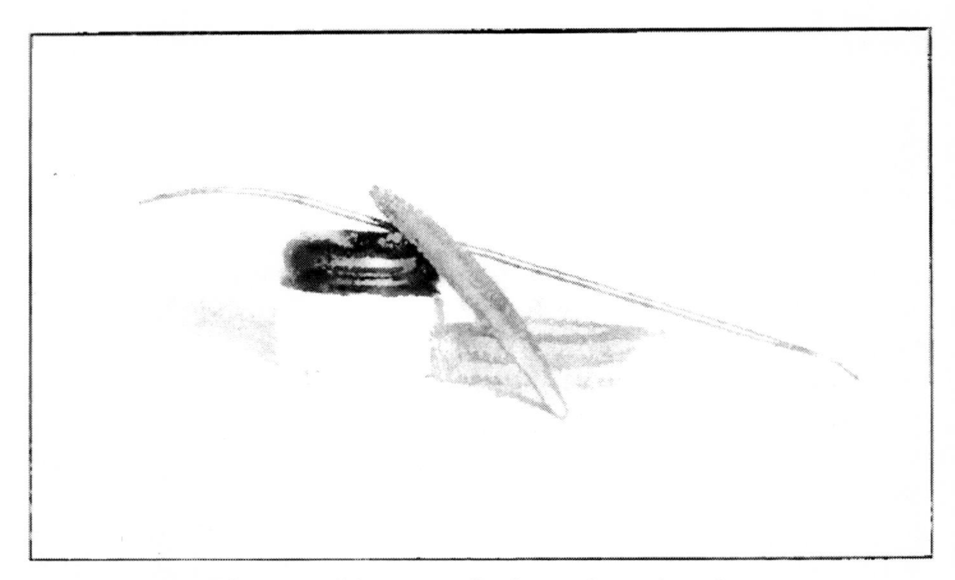

The grafting outfit is quite simple.

When no nectar is coming in, this colony should have been fed the same as the cell-finishing colonies already described; but, if even a very little nectar is coming in, no feeding will be necessary. The best results cannot be obtained by grafting hungry larvae. If they lie in the bottom of the worker-cells dry with no larval food around them, they are not fit to be used for grafting. They will not be accepted by the bees so readily nor can so good, strong queens be reared. Stunting the larva at the beginning of its development can not be overcome at a later period, no matter how ideal the conditions may be. If the larvae are floating in royal jelly, they are in perfect condition for grafting. If they are not, it indicates that the colony needs feeding. Should the colony not have a sufficient number of larvae of the right size, it is a good practice to insert occasionally an empty comb in the center of the brood-nest in which the queen may lay. If possible a black comb should be selected since the larva can be seen much better than in a new white one. Use a brush to remove the bees from the comb for, if the comb is shaken, the nectar will be scattered over the larvae, in which case they are not accepted so readily.

Carry the comb into the grafting room. Now take three bars of cells that were dipped during the winter. Be sure that the cells are perfectly clean. If they contain any dust or dirt they must be thoroughly washed and dried before being used, as the bees will not accept dirty or dusty cells. With the large end of the grafting needle place a little diluted royal jelly into each of the sixty cells. A drop about twice the size of a pinhead is sufficient. Endeavor to get this in the neat round ball right in the bottom of the queen-cell, for the bees accept them much better when it is placed in this manner. Keep the cells and the comb out of the bright sunlight as much as possible. When the weather was hot and dry, I formerly sprinkled water on the floor to keep the air moist that the jelly and larvae might not dry out and die. One day as I was doing this the Office Force was looking on in that inquisitive manner common to the gentler sex, and she said, "what is the use of dampening up the whole room when you merely want to keep those cells moist? Why don't you dip a bath towel into some water, wring it out and spread it over the cells?" "Yes, why didn't I? For the very good reason that I never thought of it. Thanks for the bright idea." I tried it. It works to perfection, so I have used it ever since. A moistened towel keeps the cells from drying out and protects the larvae from light and dust.

Now sit down in a chair with your back to the window so the light will come over the right shoulder. Place one of the cell bars on the side of the top bar of the brood-frame and parallel with it, holding it there with the thumb of the left hand. With the grafting needle in the right hand carefully slide the point under the larvae, choosing one that is about twelve hours old. Larvae that are twelve hours old are extremely small, and unless the one doing the grafting has very good eyes, he will be unable to see the clearly enough to do satisfactory work. A fine rule is to use larvae as small as can be seen; but, if the operator has exceptionally good vision, there is danger of getting them too small, for larvae less than twelve hours old are not accepted so readily as those older. I have never been able to determine whether it is due to the fact that these small larvae can not stand the handling or whether for some reason the bees do not like them so well. On the other hand, larvae much more than twelve hours old should not be used, for while they will be accepted, they often do not make such good queens as the younger ones. True, they have not yet received any food except royal jelly; but from experiments I have made, I am sure that the best queens can not generally be produced if older larvae are used. I believe it is due to the fact that, as the larva is grafted at an advanced age, the nurse bees do not have the same length of time to store royal jelly in the queen-cell as in the case of younger larvae, therefore the larva does not have sufficient food for it's fullest development.

Holding it there with the thumb of the left hand.

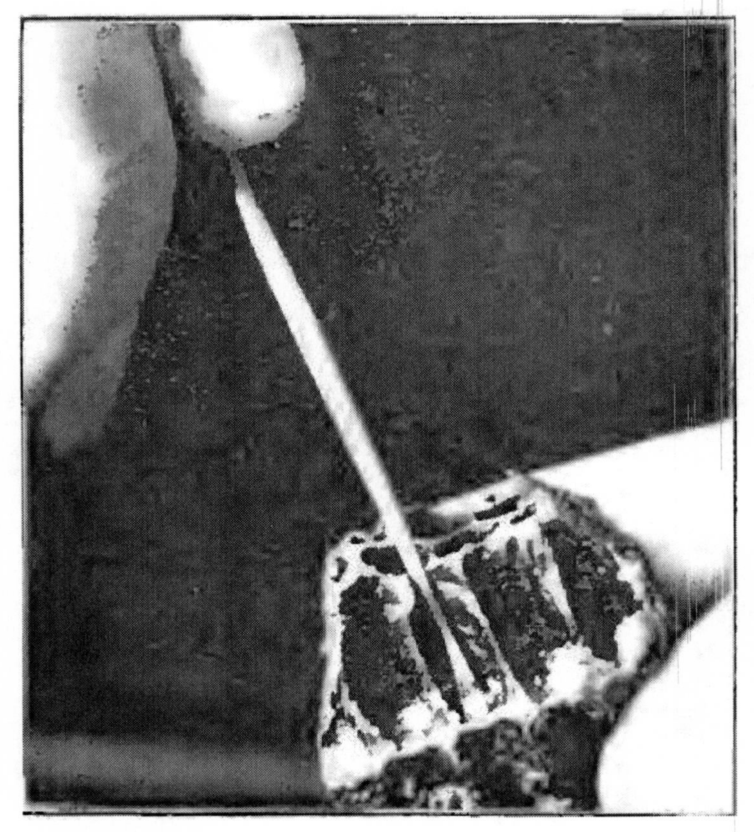

Slide the point under the larva.

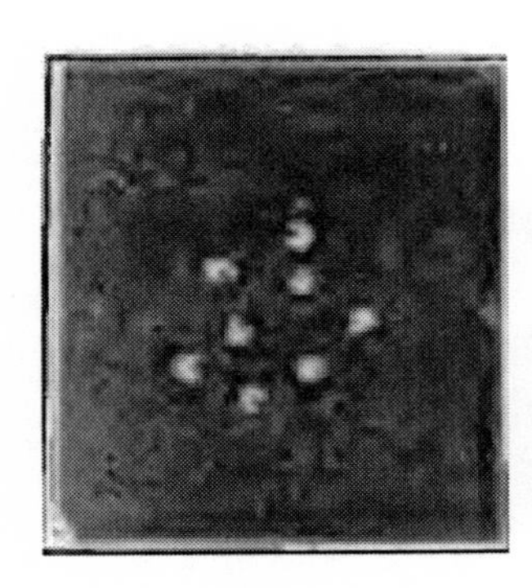

Larvae 12 hours old

One-half of the larva over the edge.

There is quite a knack in getting the larva on the grafting needle in just the right position. If it is entirely on the grafting needle, difficulty may be experienced in getting it off. The needle point should be placed under it in such a way as to leave about one-half of the larva projecting over the edge. When this is done, it is easy to remove the larva from the needle. Put the needle into the cell cup until the larva touches the drop of royal jelly, carefully draw the needle sidewise and the larva will remain in the jelly. At first this may be a slow operation, but in a short time you will be able to graft three bars in ten minutes or less. As soon as one bar is finished, place it back under the dampened towel. When all three are finished, they are ready to be placed in the swarm box. Take them to the basement and bring the swarm box out of the dungeon to the front where the light is better. Put the three cell-bars tightly together so they may be grasped with the right hand all at once. Lay them down with the cells upward. Now pick up the swarm box by taking hold of each end. Raise it about six inches from the concrete floor and bring it down with a jar. All of the bees that were hanging to the cover will fall to the bottom in a mass. Now remove the upper cover, take up the cell bars and place them in the box, allowing them to rest on the two cleats in each end, and slide the cover back in place. This can be done without a single bee's getting out. Have ready a quart Mason jar, with a perforated lid, filled with sugar syrup or honey diluted with about one-fourth water, which is much better feed for the swarm box. Sometimes bees will not take the syrup, but they will take honey. I do not put the feeder on when I first set them in the basement, for they will not take syrup or honey until they have been in there several hours, long enough to get the spilled honey all cleaned up, but feed them immediately after grafted cells are given them. Set the swarm box back into the dungeon, and the bees will do the rest.

Digression.

We have now passed over the most difficult part of queen-rearing, that of getting the cells properly accepted. Many have failed at queen-rearing because they could not get cells accepted with any degree of certainty. The question is frequently asked, "Why use the swarm box? Why not give the cells immediately to a colony?" The answer is, "Because the bees will not accept the cells." The condition brought about, as has just been described, enables the nurse bees to feed the larvae and draw out the cells in the best manner possible. Bees have many peculiar traits. One of them is that they will mechanically go ahead with a job that has been started. It is not difficult to get cells finished when they are once started. The difficulty is in the starting. So now, as we have brought about a condition by which the bees will start cells, it is a simple matter to get other colonies to go ahead with the job and rush it to completion; while, if we had given those same colonies these cells to start, they would have had none of it.

Reasons of Success.

Let me enumerate again the reasons why bees accept cells.

1. Liberal feeding of the colony that furnishes bees to stock the swarm box. Remember, unless a honey flow is on, they must be fed generously at least three days before being placed in the swarm box.
2. A sufficient number of bees in the swarm box. There should be at least five pounds.
3. A large number of nurse bees, young bees. Re-member that old bees are poor nurses and will fail in accepting cells. (Transcriber's note: Jay believed this because the scientists told him this. After his own careful observation he changed his mind on this.)

4. Well-fed, moist larvae in the colony which has the breeding queen.

5. Clean cells, made of wax that has not been scorched in melting, are most essential.

6. Cells must be the right size. Those that are too large will not be accepted.

7. Be careful to keep the royal jelly at the right consistency. Royal jelly too thick or too thin will cause failure in whole or part.

8. Grafting Larvae that are the right size and age.

9. Be careful that the larvae have not been over-heated or dried by the sun's rays. If they are, the bees will always reject them.

10. Keep the larvae from chilling.

11. Careful handling while grafting so as not to injure or kill the larvae.

12. Be sure the swarm box is kept in a place that is neither too hot nor too cold.

And lift out the bars.

The bees should remain in the swarm box until three or four o'clock of the day following. If taken out earlier the cells are not sufficiently advanced to insure their completion by the finishing colony. If left in too long, the nurse bees seem to exhaust their supply of royal jelly and the larvae are not sufficiently fed. They can be taken out at any time before the night of the following day; but as a general thing, the cells suffer if left in over night of the second day. From twenty-four to thirty hours is the proper length of time for the bees to be confined. Carry the box out to the hive from which the bees were taken, remove the upper cover and lift out the bars. If the work has been properly done nearly all of the cells should be accepted. One should average an acceptance of eighteen cells out of twenty and frequently all should be accepted. When the bars are taken out the larvae should have an abundance of royal jelly literally swimming in it, and the cells be drawn out into proper shape. If conditions are right all sixty are accepted.

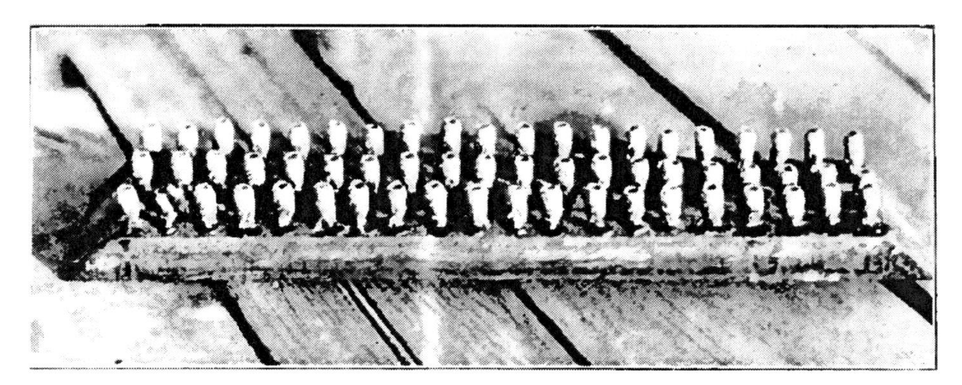

All Sixty are accepted.

Give the bar a very light shake to remove most of the bees that are clustered on it and then carefully brush off the remainder. Do not shake the bar too severely or some of the larvae will be displaced, in which case they will be removed from the cells by the finishing colonies.

For suspending the cell bars in the finishing colonies there is nothing better than a regular cell-bar holder made to hold three bars. Place one bar in the bottom section, spread the brood-frames apart, put in the holder frame, and give it to one of the finishing colonies. The best results are secured by placing frames of unsealed brood on both sides of the frame that holds the cells, for this draws the nurse bees right to the cells and they immediately take hold of the cells and carry the work on to successful completion. Replace the cover and see that the feeder is kept liberally supplied with feed so that the bees will receive an ample supply. Take the other two bars and give them to the other two finishing colonies. Go back to the swarm box, take off the lower cover, remove the combs and shake off all the bees possible and brush off the remainder. Pick up the swarm box, invert it and give it a rap on the ground to dislodge all the bees. Replace the combs, put on the covers and take it back to the basement to remain till needed again.

Chapter XIII. The Pritchard Forced Cell Starting Colony.

While I myself much prefer the swarm box for securing cells, there are others, like my friend Pritchard of the A.I. Root Company, who prefer to use queenless and broodless colonies for cell-starting. Instead of going to a colony and shaking bees into the swarm box, Mr. Pritchard prefers to reverse the process by leaving the bees in the hive, removing all the brood with the queen, and placing them temporarily on another hive. He proceeds as follows:

All the combs including the bees, brood and queen of a medium colony (not a strong one) are removed from the hive. Two combs of pollen and honey are selected and set backing the hive one on each side, taking care not to get the queen. Two frames for holding cell bars (without the cell bars) are then put in the center of the hive. The two combs of pollen and honey are shoved over next to them. On either side is put a thin division board feeder containing thick syrup. The remaining space on each side is then filled out with dummies of division boards. The two cell bar frames are now removed and all the other combs of brood and bees are shaken into the space vacated taking care not to get the queen. The brood and the queen are now put in the upper story of a strong colony, over a queen excluder. It is necessary to cage the queen.

In from half an hour to one hour's time, or as soon as the bees in the made-up colony set up a roar of distress when they have discovered their loss of a queen and brood, the two cell bar frames are supplied with prepared cells, and are then put in the open space left, where the bees are crying for a queen. The queenless and broodless bees, supplied with an enormous amount of pap which they expected to use in feeding their young larvae but which has been all removed, immediately accept and supply the prepared cells with pap. The two feeders containing thick syrup and two combs containing honey and pollen will give the bees all that is necessary to supply them with material for making more pap.

The prepared cells will usually be accepted and lavishly supplied with pap in about 24 hours. It is not advisable to remove them before. When nicely started the cells are put in cell-finishing colonies as previously explained. The removed brood and the queen are restored to the colony.

While it might and could make a second batch of cells, Mr. Prichard does not advise it.

Mr. Pritchard says he prefers this method of getting cells started, because it saves the extra equipment of swarm boxes, toting them back and forth from the cellar, and because the bees during the time that they are starting cells are not confined. He thinks this is very important. The unconfined bees, he thinks, will do more and better work. By the plan described, he says he can get 200 cells started, each cell literally gorged with a big supply of pap. These cells, when given to cell-finishing colonies, will be completed in the regulation time, and every cell will be perfect.

The fundamental difference between the forced cell-starting colony and the portable swarm box is that Mr. Pritchard moves the brood and the queen, while I move the bees. He thinks that the unconfined bees do better work. I am not so sure of that. I succeed better with the bees that are confined in a cool place. Mr. Prichard will doubtless continue his way, and doubtless I will go on with the way that has given me the results I have secured.

It will be noted that the Prichard plan of making up a cell-starting colony amounts virtually to a swarm box left on the old location. Fundamentally the *principle* is exactly the same, but the procedure is different.

Of course it is necessary to feed these prepared colonies the same as the bees in the swarm boxes. This is important.

A Modification of the Doolittle Plan.

I have used a modification of the Doolittle method with the best of success and it may be preferred by many to the swarm box. With further experimenting and practice, possibly we ourselves may prefer it. This method is as follows: The colony for starting these cells should be one of extraordinary strength, being a two-story colony. Any standard hive will do, but we use the Jumbo hive, and in preparing this we see that it has twenty Jumbo frames of brood. The queen-excluder is kept between the upper and lower story. After all of the brood above has been sealed, we remove the lower story containing the queen and the brood to a location about ten feet distant. The upper story, containing only frames of honey and capped brood, is set on the bottom-board of the stand from which the hive containing the queen and brood was removed. Three frames are taken out to make room for three frames of cells. The hive containing the queen is now opened and the frame containing the queen is set outside. The frames are then lifted out and the bees from eight or nine of them shaken into the hive on the old stand. The queen with her frame of brood, is then set back into the hive on the new stand and the cover replaced. In a short time, half an hour or an hour, as soon as the bees have cleaned up the honey or syrup that has been daubed on them, they are ready to receive the cells. Three bars are grafted, placed in a frame to hold them and put into the starting colony. A second and third are prepared in the same way, which fills up the space. In this manner nine bars are started instead of three as with the swarm box; though it is not advisable to start so many unless the colony is of tremendous strength. After twenty-four or thirty-six hours these frames are removed and given to the finishing colonies, the same as when the swarm box is used. The other brood-nest containing the queen and brood is now brought back, set underneath and all is well. In two days' time the colony can be used to start another batch of cells. We use the same colony over and over for starting cells twice a week.

By this time the objection to the original queenless, broodless method is largely overcome. The queen can continue laying, as enough bees are left with her to keep up brood-rearing, and the work is much less than with the method whereby the queen has to be caged and bees brushed off from the brood. In using this, if there is no honey flow they must be fed the same as when the swarm box is used. Some will prefer this to the swarm box, and some will not. It has one advantage over the swarm box, in that no harm is done; while some have reported bees suffocated in the swarm box. Another time-saving feature is that if this colony is made tremendously strong as described, a larger number of cells may be started.

To say that a colony must be strong does not mean much; but if both stories are kept full of capped brood before being used, it will build the colony up to greater strength; and if it should run down, the upper story may again be filled with brood. But this should not be left in this story when giving cells until the brood is sealed, for the bees will not do so good work at cell starting where they have other brood to feed. Recently we had a field meeting in our apiary in Vincennes. Always willing to do their part, my bees decided it would be a good exhibition of they would swarm; so one of these cell-building colonies did so just before the crowd arrived. I put a piece of burlap on a saw-horse and placed the queen on top, and the swarm settled there where it remained for two hours or more, greatly admired by those present. Several disinterested parties weighted this swarm and found it weighted exactly twenty-five pounds. This would be about one hundred and twenty-five thousand bees, while seventy-five thousand must have still remained in the hive, thus making the enormous horde of two hundred thousand bees! I had estimated our starting colonies and finishing colonies at one hundred and fifty thousand each, but from that experience, I believe my estimate was too low. Of course, it is understood these bees were not the product of one queen as it would be impossible for one queen, no matter how prolific, to produce a swarm of that size; but from this we get some idea of a really strong colony, and for best results in raising the finest cells in large numbers, a strong colony as just described is very essential.

In the above modification of the queenless, broodless method it will be seen that the brood and queen remain together, and the queen continues to lay very similarly to the method in which the swarm box is used. The cost of equipment of the two methods is about the same. One requires the swarm boxes and funnel and the other requires an extra hive. Personal preference must determine which is used; both are excellent.

Extensive experiments are being continually conducted at our yard, and possibly in the future the present plan of cell-finishing above the excluder will be abandoned for something better. We have left one hundred and eighty cells with the queenless, broodless colony that started them until they were completed and some most excellent results have been obtained. Two colonies were shaken into one, and both colonies moved to a new location. Further experiments will determine whether or not this method will supersede the former with us. In any of the above methods, however the one big feature is proper feeding. Heavy feeding three or four days before giving the cells is a most important feature.

Chapter XIV. Our Daily Program.

Many will find it more suitable to their systems of management to graft three or more times a week. Some graft every day. I graft Wednesdays and Saturdays. As I am to give my own system first, I shall follow it through. Let us consider that we did our first grafting on Saturday. Before time to dispose of the cells we shall again graft on Wednesday and Saturday. To make it clear, let us suppose the Saturday we first grafted was the 5th of the month. We must graft again on Wednesday, the 9th, again on Saturday, the 12th. When we put in the cells of the second grafting, some of the cells of the first will be capped over. The bar containing these should be moved up to the middle and the new one placed on the bottom shelf. The bees do better work when the newly accepted cells are placed on the bottom shelf. By this method, the bar of ripe cells is always on top and it will not be necessary to remove the frame to get the bar. Cells should be removed and introduced to colonies or nuclei on the 10th day after grafting. So the cells that were grafted on the 5th must be introduced on the 15th. This makes Saturday the busiest day of the week as we must both introduce cells and graft, but the light day's work comes on Sunday so we can enjoy a day of rest. While we try to arrange our work so there is as little work as possible on Sunday, there are a few chores that cannot well be avoided. Laying queens are removed from the nuclei on Mondays and Fridays so that by following this program everything fits nicely. All that there is to remember is the following program:

Program for the Week.

Monday: Remove laying queens from the nuclei.
Tuesday: Introduce ripe cells to nuclei.

Wednesday: Graft.
Thursday: Empty the swarm boxes.
Friday: Remove laying queens from nuclei.
Saturday: Introduce ripe cells to nuclei and graft.
Sunday: Empty the swarm boxes.

This program prevents mistakes, and, while no record or memoranda are used, it is almost automatic in its working.

In explaining this program at this time I am getting a little ahead of my story, but it seems necessary and details will be explained in due time. One feature must be watched carefully. In the finishing colonies queen-cells will sometimes be started on the frames of brood. If any queens are allowed to emerge, havoc will be wrought with all three bars of cells. It is well each time a bar is placed in the finishing colony to look the brood-frames over and cut out any cells that may have been started.

Chapter XV. Nucleus Hives

There are many styles of nucleus hives in use, and some have desirable features not found in the others. The small Baby Nucleus hive had a run for a while but is now generally considered a mere passing fad. It is so small that the bees are put into an unnatural condition, and they therefore perform in an unnatural manner. They seem to delight in pulling off all sorts of crazy stunts, such as absconding with a laying queen or absconding with a virgin; absconding when they run out of food or absconding when they have plenty. Another of their favorite sports is balling their queen when she returns from her mating flight. I have seen queens fly out from their baby nucleus and, unlike Lot's wife, they never looked behind them. The queens reasoned, "Why take a look at that little hive? I'm not coming back!" And frequently they did not, but would hunt around trying to find a real colony that would accept them. I used to keep a number of bars in a single hive for incubation. These colonies seemed to be particularly inviting to these truant queens, which were usually accepted. There was henceforth a great tearing down of cells, and it made me very dejected to see a double handful of dead queens lying in front of these hives. They never worked this game, however, where cells were finished above the excluder. These baby nuclei are easily robbed out, do not gather enough to live on and do not stand either hot or cold weather as well as the larger ones.

I have a vision of one day during a dearth of pasture in hot July when a baby nucleus absconded and went up into a tall tree and clustered. Was it worth going after? Maybe they had a laying queen, so I would try. The whole swarm was not much bigger than a walnut, shucks and all. At last, after climbing till I was completely tired out and had almost reached the, they took wing. While I was watching them disappear into the blue sky and was in a state of mind unnecessary to describe, along came a cheerful idiot who asked, "Say, Mister, how much honey did they make up there?"

Another objection to the baby nuclei is the fact that it is hard to tell a good queen from a poor one, for a good queen lays several eggs in one cell for want of room, exactly like a poor queen. Baby nuclei do not contain sufficient bees to incubate the queen-cells properly, thereby resulting in inferior queens. Yes, I have in use a hundred baby nuclei-as playthings for the children and for use as bird boxes. A woodpecker has appropriated one and, after peeking the entrance a little larger, crowded it full of acorns. For once the baby nucleus has secured a surplus.

The Root twin mating nucleus hive is midway between the Baby and the ones having standard frame. In it each compartment holds three frames, just the size for three to fit inside a regular Hoffman frame. These may be placed in a regular hive, and when the frames are filled with brood and honey they are taken out and placed in the nucleus hive. J.E. Wing, the well-known queen-breeder of San Jose California, prefers this hive to any other. He has special hives made to hold large numbers of these frames, for stocking those for the nucleus hives. In this way he overcomes one objectionable feature, that of fitting them into a regular Hoffman frame. Mr. Wing's system of management is favorable to these small hives for he practices migratory queen-rearing, moving to localities where there is a honey flow. He moves to one district where there is a heavy flow from honeydew. It should be remembered that the small nucleus hives give much better results when there is a honey flow than when there is a dearth of pasture so that feeding becomes necessary. The large "babies" give some better results than the smaller ones; but the ones taking a regular brood frame have so many advantages that they are being used by nearly all who rear many queens.

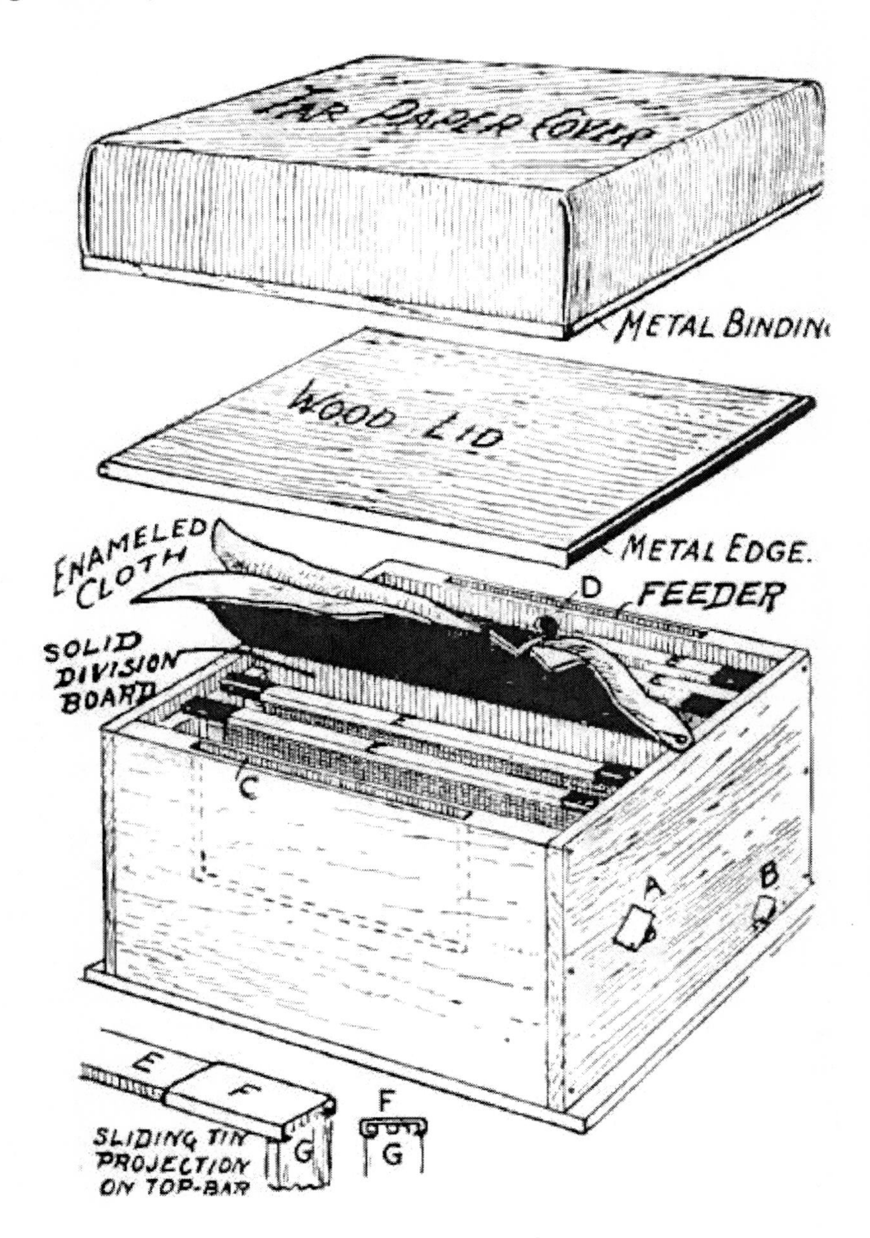

Root twin mating nucleus.

Each compartment large enough to hold two Jumbo Frames.

I strongly advise a nucleus hive that will take the regular brood-frame that is used in your hives. The one that I use is a twin hive, each compartment large enough to hold two jumbo frames and a division-board. The entrances are at opposite corners. A telescope cap is used with cleats that support it and give the air a chance to circulate, thus keeping it cool in hot weather. Usually only one frame is used with a thin division-board feeder to keep the bees from building comb in the vacant space. This gives the bees so much room that there is no absconding. It is comfortable in the hottest weather and has given perfect results. During a honey flow these nuclei are strong enough to fill up with honey. In fact in many cases, I have to give them sheets of foundation to keep the bees from going over the feeder and building comb.

Another nucleus hive among the best is that used by M.H. Mendleson, the veteran honey producer and queen-breeder of Ventura, California. This is a standard hive body, divided into three compartments. There is an entrance at each end, and one on one side. While working with this hive Mr. Mendleson sits on the blind side. In their mild climate he is able to winter these nuclei over with perfect success. There is nothing better than this style of nucleus hive.

Entrance Blocks for Nuclei.

The entrance block may seem like a trifling item; but, after experimenting with several, I feel that a description of one that has given me splendid satisfaction may be worth while. If the entrance is not very plain and easy to enter, the queen on her return to the hive will have difficulty in locating it readily. I have witnessed queens returning many times, and when I used an inferior type of block, the queen, after trying in vain to find the entrance, would go to another nucleus and try there.

The one that has given me perfect satisfaction is made to slope towards the entrance so that, if a bee alights within a few inches of it, the block guides her directly in. When the young bees fly out for the first time, they have no trouble in finding the entrance at once. Three nails are driven through the block from the upper side so that the points are barely exposed. When it is desired to confine the bees, as is necessary when forming nuclei, the block is placed over the entrance and pressed down. The nail-points catch in the wood so that the block cannot be pushed away by the bees. When they are first released, this may be moved enough to give an entrance just large enough for one bee to pass, and later it may be moved to allow the full entrance. In this small opening a single bee will stand guard and is able to keep off all oncomers in a manner similar to that of the Spartans at the Pass of Thermopylae.

Shade.

Shade is a very important item with nuclei. This is true in a greater or less degree, depending upon the style of cover in use. We use the deep telescope cover which shades nearly the entire side of the nucleus. In addition, the cover has an inner lining of thin boards with cleats on both sides so that there is a double air space, one above this inner cover and one below. Very little difference is noted in the behavior of the bees in such nuclei whether they are placed in the sunshine or in the shade. On the hottest days, however, the bees cluster out less where in the shade.

In case a single cover is used, shade is a necessity. Years ago when we used the baby nuclei, some were in the sunshine and many cells did not hatch; and when they did, many of the virgins were small, dark and inferior. A grove is the best shade, and if the trees are far apart to admit the sun in spots, it is all the better, for on dark days one can step out into these lighter spots to examine the combs for eggs, etc. In case the thermometer does not go above ninety degrees, the telescope cover with two dead air spaces gives perfect results.

Achord queen rearing apiary.

Chapter XVI. Virgin or Cell Introduction.

A number of methods are used to get the virgin queen safely into the nucleus in which she is to lay after mating. One is by allowing the queen in the cell which is kept in a strong colony for incubation to emerge in a nursery cage. I used that method for several years, but have discarded it since I could not get so good, vigorous queens in that manner. I found there are two reasons for this. One is because the bees are unable to cluster closely around the cells in order to keep the temperature right, and the result is faulty incubation of the tender pupa. This defect manifests itself in two ways, by smaller queens and darker queens. If the cells are kept too cold, it makes the queens dark. Sometimes in early spring the cells were a little below the proper temperature, in which case no harmful effect was noted but the queens were darker in color. When they emerged in the nursery cages and the weather was cold they were both dark and small, and a number of the cells did not hatch. Now, in theory, if the cells are placed in an upper story over a strong colony, the temperature will be the same as though no cage is used. But if the bees are ventilating the hive or fanning to evaporate the nectar or syrup, a current of air is blowing through the hive. If the bare cell is in the hive the bees cluster tightly around it, thus protecting it from air currents and keeping it at just the right temperature; but if the cell is in a cage, the bees pay no attention to it whatever, so that the cool current of air blows right though the cage and chills the cell.

The second reason why too many inferior queens are reared when the nursery cage is used, is because the new virgin does not receive the proper feeding when she first emerges, and this at the time that she most needs abundance of food of the right kind. When she emerges in the cage she has to depend upon the candy that is placed there. Sometimes she eats enough of this to keep her alive, and sometimes she perishes from an enforced hunger strike. Sometimes the bees feed the queens through the wires, and sometimes they do not. I have noticed some peculiar traits of the bees in this connection. They occasionally cluster around one cage and give that virgin all the attention in their power while they are balling another cage, and probably the rest are ignored altogether. I presume they had agreed to accept that one queen as their own and let all the rest go hang.

That the virgin does not receive the proper food and care while in the nursery cage and that her development is retarded, I have proved many times. As we know, seven to eight days usually pass from the time a virgin emerges from the cell until her mating flight. I noticed in many cases that a longer period elapsed before the virgin mated. Usually, in using the nursery cage, the virgin mated eight days from the time she was released from the cage, that is, the virgin remained in the cage three days; then it was eleven days from the time she emerged from the cell till she mated. The time spent in the nursery cage seems time lost as far as development is concerned. In many cases these queens never turned out to be first-class queens. When they are released from the nursery cage as soon as they emerge, not so much damage is done, but even then they frequently suffer on account of improper incubation at the time that Nature is putting the finishing touch on the pupa. Some had legs that were paralyzed or withered and wings underdeveloped.

Moreover, I noticed that the young virgin seemed to like the dried-up royal jelly that remained in the cell from which she emerged. She would eat it all, notwithstanding the fact that she had nice candy made out of powdered sugar and honey. Sometimes the queen would burrow into the bottom of the queen-cell and die there. From these observations I reasoned that the virgin needs royal jelly, and the only satisfactory way for her to get it is to have it fed to her by the nurse bees. I then conducted some experiments along this line. A number of cells were caged, and a number from the same grafting were given direct to nuclei. Those emerging in the nursery cages were introduced to nuclei in the same cage from which they emerged, taking about three days for the introduction. All from the cell-introduced lot with one exception were laying before the first from the cage-introduced lot. A number of experiments of similar nature were carried on, and all showed conclusively that the queens were injured by remaining in the cages at this time. So we can lay it down as a safe rule that in order to get the best results the virgin must emerge among nurse bees in order that they may give her the proper food and care. In an article in the American Bee Journal, I gave my experience with the nursery cage, and in a footnote Editor C.P. Dadant stated, "The editor applauds with both hands at these conclusions, for he had also tried the queen nursery years ago and did not like it." With the backing of such an acknowledged authority as Mr. Dadant, I fee sure of my ground in this connection.

The Cell Protector a Hindrance.

The cell protector was discarded for the same reason. The bees cannot properly care for the cell when it is in the protector. However, this is not so noticeable as in the case where the cell is caged; but there is little, if anything, to be gained by using the protector. It is a known fact that bees will accept a cell much more readily than they will accept even a newly emerged virgin. That being true, if they would not accept an unprotected cell, they would not accept the virgin when she emerged. In fact, from several experiments I am convinced that the cell protector is a hindrance rather than a help. At one time we introduced on hundred bare cells and one hundred with protectors. When we came to look for the virgins, we found about 30 per cent more in the nuclei where the bare cells had been given. Of course, they did not tear down the protected cells, but they killed the virgins as soon as they emerged.

The unemerged queens at this age are very tender and should be handled with the greatest care and should not be away from the bees longer than is absolutely necessary. A number of years ago I used to lay the cells on their sides in a box containing cotton batting. I found that, if they were left in this box for any length of time, many queens emerged from them would be crippled. Their legs and wings in particular would suffer. In conversing with Mr. Snodgrass of the Bureau of Entomology, Washington, D.C., he informed me that, if the pupa lay on its side for any length of time, the circulation stopped, which results in injury to the parts affected. However if the virgin will emerge within twenty-four hours, no harm will come if a cell is laid on its side for a short time. Keep covered from cool air or hot sun and by careful handling and maintaining as nearly as possible the temperature of the bee cluster it's perfect development is quite certain.

Fig. 1.-Larvae just as they came from the swarm box.

Fig. 2.-Cells are flooded with royal jelly.

Fig. 3.-Just as the bees were capping the queen cells.

Fig. 4.-Day after cell was capped.

Fig. 5.-Changing into a pupa.

Fig. 6.-Royal jelly is dry and brown.

This series of photographs shows the development of the queen from the grafting up to the time when the queen emerges. In Figure 1 you will note that the cells are remodeled to suit the bees. Wax was added by the bees, showing they were secreting it profusely. The bottom of each cell is covered with royal jelly. Thus in the twenty-four hours of confinement in the swarm box, the bees gave the larvae the proper send-off. In order to give a better view of the larva one cell was torn open.

Figure 2 pictures the cells one day later, after they had been in the finishing colony for twenty-four hours. The marvelous growth of the larvae in this length of time will be noted; but when we notice how the cells are flooded with royal jelly, the larvae really have no excuse for not growing. Figure 3 was taken just as the bees were capping the queen-cells. The larvae are getting too fat to curl up in them any more, so are beginning to lengthen out a trifle. The two end cells are already capped. Figure 4 shows the larva the day after the cell was capped. The royal jelly is still white and soft and would be in good condition to use in grafting, if thinned slightly with water. In figure 5 we see the larva changing into a pupa. The royal jelly is drying up and getting darker in color.

The next stop in the formation of the queen is shown in Figure 6. The change from the larva into the pupa is so very rapid as to seem marvelous. This takes place in twenty-four hours and in that short time, head, legs and short wings are formed so it appears a fully developed queen with the exception of wings and color. At first this pupa is exceedingly soft. While handling one, I accidentally dropped it on my foot. It splattered out much like a drip of clabbered milk, and no form of the pupa could be found. The pupa remains in this form with very little change, as far as appearances are concerned, for about seven days; but it becomes firmer and harder continually during that period. This picture was taken the tenth day after grafting, therefore one can see the condition of the pupa at the time the cells are to be handled. The royal jelly is dry and brown as shown in the top of the cell where it dried, leaving a space in the top of the cell. Where it is found necessary to handle the cell containing the tender pupa any time before the tenth day after grafting, great care must be exercised in handling it, as mentioned elsewhere, or crippled queens will result.

When the handling of cells before the tenth day after grafting is necessary, they should be placed in holes in a block to keep them right side up. As we handle them only on the tenth day, such care is not necessary. We place a cushion in the right end of the hive-seat, on which the cells are laid. A cover is tacked on it in such a way as to keep off the sun's rays and yet be easily raised when getting the cells Crippled queens are practically unknown to us since using this method.

Can be easily raised when getting the cells.

Chapter XVII. Cell Introduction

To establish nuclei and introduce cells to them is our next step in queen-rearing. Let us consider the subject a little. Two features in queen rearing have always been difficult; first getting the newly grafted cells accepted by the bees; and second, some safe and satisfactory method of installing the queen-cell or virgin in the nucleus.

This first difficulty, the manner of getting cells accepted, has already been described, and by carefully following the directions given one should have no trouble in having excellent cells built in abundance, so that the largest and best only are kept; but when all goes as it should, every cell will be so abundantly supplied with royal jelly that there is little choice. Having mastered this point in queen-rearing, let us now pass to the feature that has in the past been extremely difficult, and at certain times of the year has seemed insurmountable.

However, few things on earth are impossible if we know the underlying principles, find the cause of failure and apply the remedy. In cases of difficulty in any walk of life, even queen-rearing, a splendid motto to adopt is, "There is a way." When things do not go as we would have them, think of this motto, and proceed to find the way. There must surely be a way in this case, I figured.

In the past we have been led to understand that the reason why bees tear down cells given them is due to the fact that the cells are strange, with an odor different from that of the cells reared in the colony. Recognizing this, the bees proceed to rear them down. Now many things pointed to this belief. For instance when a cell was given to a colony, they would tear it down while at the same time they were building cells of their own; and many times when they had cells of their own, they would leave them unharmed and immediately tear down any strange cell given them. The natural deduction, therefore, was that they destroyed the cell given them because it was not their own. If such were the case, we could expect little relief, and our motto, "there is a way," would not apply.

Every spring when establishing nuclei there was a regular epidemic of cell-destroying. They would tear down cells as fast as given to them until they ran into laying workers. If virgins were introduced to them a little better progress was made, where we used our Push-in Cage. However, I was thoroughly convinced if the very best queens were to be reared, the virgin must not only emerge among the bees but these bees must be *anxious to receive* that queen in order that they may not injure her, but on the contrary, receive her gladly with out-stretched tongues and feed her abundantly that she may develop into a large prolific queen, the kind we all should strive to produce if we are to get big yields of honey.

So the question of how to get the bees to do our will in this respect was a puzzler. It was not practical to introduce the cell in a cage as we would a queen. If what we had been taught was true, that it was because it was a strange cell, there seemed little hope of overcoming this trait of bee nature; and to set about to overcome this rule seemed very difficult since there apparently was nothing to vie us a clue. There seemed to be no starting point.

In the spring of 1922, although there was a light stimulative honey flow, the bees were the worst at tearing down cells in all my beekeeping experience. I had planned to put out six hundred nuclei in two weeks, forming one hundred and fifty at a time to conform to our program. When the first one hundred and fifty were put out, nearly all the cells given to them at the time of forming the nuclei were destroyed. The second batch of cells suffered a similar fate, and there was no use in forming more nuclei as the first would not accept cells. The mystery as to why they tore down the cells was very impressive, to say the least.

I sat down under a tree to see if I could think out "the way." I carefully went over the experiences of the past fifteen years and finally the truth began slowly to unravel before me. I remembered that upon numerous occasions I had taken out cells from one of the finishing colonies, and having had no immediate use for them, had returned to another finishing colony, paying no attention about putting them back in the one from which they were taken. I recalled that, during those years, hundreds of bars of cells had been placed back into finishing colonies strange to them, and yet in all that time, to my knowledge, *not one single cell had ever been destroyed.* Evidently, then the reason the bees destroyed cells was not because the cells were strange. What, then, *was* the reason?

What was there about those strong colonies that had a prolific laying queen confined to the lower story that caused them to accept strange cells without question, while the nuclei would tear them down as fast as given them? Why did a strong queenless colony that had been used to finish cells accept a dozen bars of cells and never tear down a single one? Plainly, the fact that it was a strange cell had little or nothing to do with it. *Plainly, it was the condition of the colony.*

If this was true, what were those conditions and could they be duplicated in the nuclei? I became quite excited over the proposition. The only difference in the condition of the finishing colony and that of the nuclei was that the colony was being fed liberally while the nuclei were not; for even when artificial feeding was not resort-ed to, the strong colony was gathering enough from the fields to bring abundance to the colony, while the nuclei, being weaker in field bees, were not getting enough for the fields to supply them and were consequently drawing on the stores present in the combs. Could it be that the secret of successful cell acceptance was merely a matter of feed? I would soon find out.

I went to a number of colonies from which I expected to take brood for forming nuclei and gave them ten pounds of thick sugar syrup late in the evening, repeating it the second night, and you can imagine with what eagerness I awaited the results. I then formed nuclei in exactly the same manner as before. Presto! *Practically every cell was accepted!*

In my previous attempts at introducing cells to nuclei, out of one hundred, eight-nine cells were destroyed; while after feeding, only four were destroyed out of one hundred. Yes, "There is a way." To be sure that the conditions as regards the honey flow had not changed, I then formed some nuclei without feeding and some with feeding the results obtained were in every way similar to those of my first experiment.

My next step was to see if this would be as entirely successful with the established nuclei in getting them to accept cells. To our great delight we found this also worked to perfection. When ready to give a cell, we took a two-quart Mason jar with a perforated lid and shook about half a hint of syrup into the combs. We found there was practically no loss of cells; and that it was not necessary to have any honey in the frames when forming nuclei. We found that another admirable feature resulting from this feeding of the nuclei is that the bees are put in a condition to give the newly emerged virgin the very best of care. As a consequence they feed her as soon as she emerges so that she develops as rapidly as though reared in a populous colony during swarming time. Virgins thus reared mature, mate and thus begin laying one or two days sooner than in nuclei where no feeding is practiced.

This method of feeding to prevent cell-destruction is practiced at any time during the year when trouble is experienced in cell-acceptance. After a nucleus is established and a good honey flow is on, feeding is discontinued. Even during a honey flow, when a nucleus is first established, feeding is necessary. In this case the fielders return to their old house, leaving a lot of emerging brood. The bees left in the nucleus can get no honey from the field until they are older; therefore the nucleus bees are hungry, and will destroy cells in their desperation. Feeding prevents this. Our nuclei are now equipped with division-board feeders, having a capacity of about one quart. These are filled with syrup the day before giving them the ripe queen-cell. Seldom indeed is a cell torn down.

Chapter XVIII. Why Nuclei Tear Down Cells.

Now that we have worked out this system whereby the cells are accepted in a satisfactory manner, let us ask the bees some questions to find out why they do things just as they do.

In the first place, when do they have cells in their colonies if left to their own devices? The greatest number of cells is found when the colony is preparing to swarm. And what are the conditions within the hive at this time? While there are a number of conditions that we cannot well duplicate in the nucleus, such as lots of bees and brood in all stages, yet there is one important feature-*plenty of feed.* Some nectar is coming in from the fields, the nurse bees are feeding the larvae, and all bees have an abundance of food. When in this condition they not only build cells of their own but will tolerate other cells if given to them. When preparing to swarm, the bees with ripe cells of their own will never tear down strange cells given them; but, let a rain come up and the weather turn cold so that no nectar comes in from the field, they begin tearing down cells whether their own or strange ones.

Now notice one point very carefully. They tear down *the more mature cells first.* Their instinct seems to lead them to realize that, if they tear down *all* the cells, they would have to start all over again and build from the bottom up, in case the weather turned warm, and nectar again came in so that they wish to swarm. If only the advanced ones were torn down, they would be retarded merely a few days. This occurs only occasionally during swarming season, but is almost invariably the rule when bees are superseding their queen. I have noticed many times during supersedure that the advanced cells are destroyed while new ones are still being started. The weather or nectar secretion seems to influence them in this respect.

This instinct, therefore, explains why, when we gave cells to nuclei, they tore ours down while at the same time they were building cells. They tore ours down because ours were advanced, and not because they were strange cells. If you wait until a nucleus has ripe cells of it's own, you can give it a strange ripe cell, and the bees will accept it without question. Many times when conditions are unfavorable, finishing colonies will tear down the cells they themselves have built. A heavy feeding will stop this destruction.

Therefore, we can accept as a rule of the bees that they will not tolerate cells when they are hungry; but, if they are lavishly fed either from a natural honey flow or by receiving sugar syrup, they will tolerate and accept cells. Moreover, there are some less important conditions that render cell acceptance certain. There should be capped brood in the nuclei and, if possible, brood in all stages; but, if these conditions are not present, a heavy feed will offset the lack of the former to a high degree. When there is no brood, cell acceptance is more uncertain.

You ask, "How about laying workers?" Well, we were talking about the laws that govern bees, and these do not apply to laying workers as the laying worker is a Bee Bolshevik and knows no law. However, a Bolshevik is more amiable on a full stomach; so some laying workers can be made to get into line by feeding, but it is better to give them a frame of emerging brood, after which a hearty feed gets them ready to accept the cell.

If this idea of feeding is applied to the different phases of beekeeping, the same results may be obtained during a dearth of pasture as during a honey flow. Many have observed how much more successful queen introduction is during a honey flow than at other times. Since discovering that feeding prevents cell-destruction, I have recommended to those having trouble introducing queens that they feed the colony heavily while the cage is in the hive. Several have reported that this gives as good results as they can get during a honey flow.

Some time ago the honey method of queen introduction was advocated. The plan was to daub the queen with honey and run her in. Some reported success; others, failure. Those who poured a pint or more of honey on the bees and the queen had better results. Now, it was not the daubing of the queen that helped; it was *the feeding*. In other words, for successful introduction, duplicate a honey flow by heavy feeding. The feeding should be done at night to prevent robbing.

In feeding with the division-board feeder it is necessary to keep the nuclei in strong condition, for robbers are always hanging around ready to pounce upon the nuclei if opportunity offers. By providing ventilation in the bottom of the nucleus hive and contracting the entrance to one beespace, and keeping it strong in bees, no serious trouble is caused by robbers. Sometimes, when they are exceptionally bad, we close the entrance entirely. In this way all the robbers that get on the combs are shut in the nuclei and can not go back home to spread the news that free plunder is to be had. Consequently, other robbers are not sent out to hunt the source of supply and marauding expeditions are restricted. After fifty or twenty minutes, the entrances are opened, when the robbers that were trapped rush for home. The nuclei are by this time reorganized and able to stand off all oncomers. In feeding either nuclei or cell-building colonies, it is necessary that they have some empty comb in which to deposit the food, for if all available space is occupied by stores they will fill up on this and become lazy and are easily robbed out.

Let us now take up our program where we left off. Before forming nuclei, we should have the nucleus hives in their places on the stands where they are to remain through the season. Ours are placed in rows running between the colonies, which are set four in a group for wintering in the quadruple cases. There is a big advantage to having the nuclei among the colonies, for much time is saved in drawing brood from the colonies, getting cells from the finishing colonies, etc. Again, when the virgins take their mating flight they have to run a gauntlet of drones, and mating is made certain. In addition all virgins are mated to your own drones, eliminating any chances of mating to drones in nearby apiaries or with drones from colonies in trees.

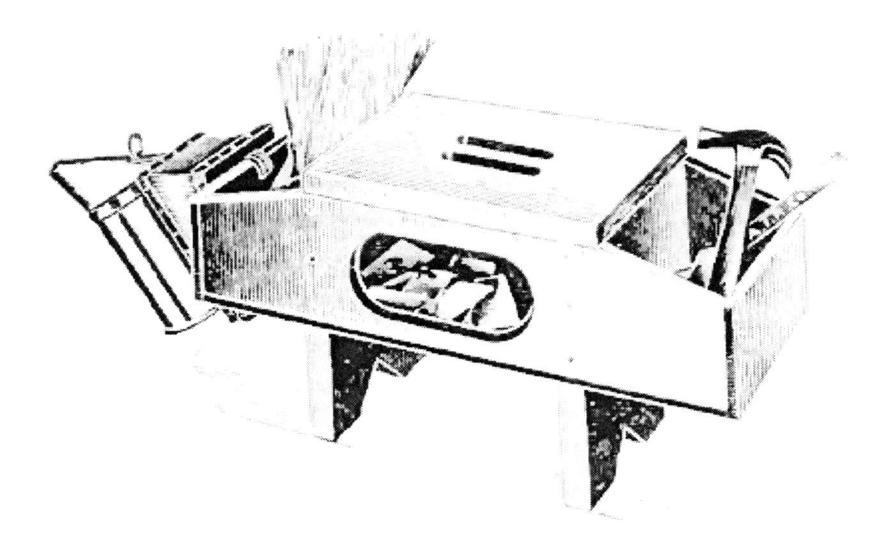

Hive-seat to accommodate the necessary equipment.

I do not believe queens go so far to mate as is generally supposed. I believe that Nature intended that the virgin should mate with drones from her own colony, for you will notice that bees never kill off their drones when they have a virgin in their hive. In their natural state, where bees are in trees, in many cases they are from half a mile to two miles from their nearest neighbor. What chance would a virgin have in mating with drones at such distances? Usually in the afternoon the young bees, the virgin queen and the drones all come out for exercise, and while circling within one or two hundred feet of the hive the virgin mates. Nature has put a check on injurious inbreeding, in that the drone that mates with the queen immediately dies, and if there should be an after-swarm the accompanying virgin would mate with a different drone.

The comb-box one of the handiest conveniences.

Formerly, when forming nuclei, we used to place a number on a wheelbarrow and take them to the hives, fill them and then set them on their stands. I prefer now to place them on their stands and take frames of brood to them. Right here let me mention two articles of yard furniture quite necessary to the comfort of the queen-breeder-the hive-seat and the comb-box. One should have a hive-seat large enough to accommodate the necessary equipment. In addition I certainly recommend the comb-box, one of the handiest conveniences about any apiary. This box is made of half-inch lumber and is large enough to hold seven frames. It has a bee-tight cover and can be used to store empty combs or frames of honey and to carry frames of brood and bees to form nuclei or put into the finishing colonies.

Into each nucleus, as it rests upon the stand where it is to be for the summer, we put a narrow division board feeder, which holds about one quart of syrup when full. We have found these much better than the two quart Mason jar with perforated lid, which we first used, for, while these jars give good results during a robbing season the robbers are apt to become a nuisance to the nuclei. This nucleus feeder measures 3/8 inch, inside measurement, is made out of ¼-inch material, and is the regulation size of a division-board. This feeder should be filled the day before a cell is given, and when this plan is followed there is practically no loss of cells.

Two days before time to form nuclei and introduce our first cells, the colonies from which the frames of brood are to be taken should be fed liberally. Any feeder will do, but I prefer to fill their bottom-board feeders at night for two nights. This is necessary to put the bees in condition to accept cells upon their arrival in the nuclei.

It is now the fifteenth, the time for introducing the cells from our first grafting. The comb-boxes are taken to the colonies that have been thus fed, and frames containing brood and some honey with the adhering bees are put into them. All frames should have considerable capped brood. Into each nucleus put one frame with its adhering bees, fill the feeder with one quart of thick syrup, and close the entrance.

Then go to one of the finishing colonies, take out the bar of ripe cells, and with a sharp, thin knife cut between the cells in case these have been built together, web-footed fashion. Then run the knife under the base of one, lift it off, put it very gently into the box in the hive seat, being careful not to invert or jar it in any way, and remove all the cells in like manner. Now go to the first nucleus hive to receive a cell. Take one from the box, lift out the frame and gently press the cell into the combs just above the brood. Replace the comb so the cell will come next to the partition which separates into the two nuclei, and thus be in a warmer place. Do the same with the other compartment of the nucleus hive, and replace the cover. Continue thus until your supply of cells is exhausted. Close the entrance tight so no bee can get out. In the bottom of the nucleus hive at the back end is a hole about one inch in diameter in the bottom-board. The hole is covered with coarse screen cloth, and left open all the time. This is quite necessary for, when the bees are confined, it affords them ventilation. In hot weather when robbers are troublesome, this enables one to close the entrance when the bees will get abundant ventilation through this opening.

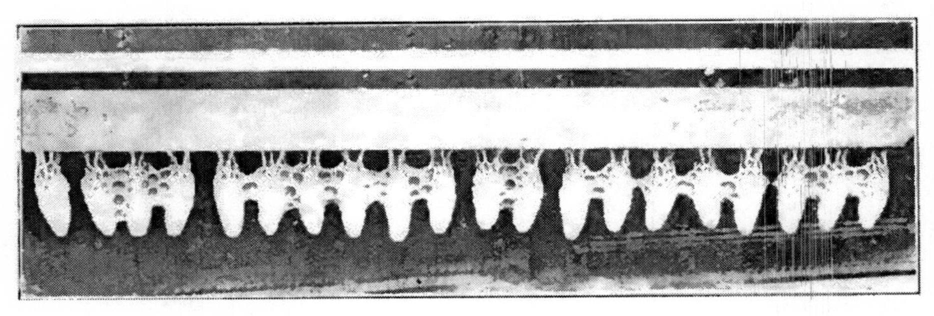

Built together web-footed fashion.

Leave the bees in these nuclei confined two nights, but liberate them after dark on the third night. By this time their old home is forgotten, and most of them will remain in this new location. A few of the old ones will probably go back, but not enough will desert to do any damage.

By following this method carefully, practically all cells should be accepted. If any are not, it is important to see that the feeder is filled when giving the second cell; for no matter how much nectar is in the fields, too few old bees are in the nucleus to gather it, and, when the young bees emerge, they will require lots of feed. If not abundantly fed at this time, they will tear down nearly all cells.

The day following cell introduction the virgin should emerge, and all is well. The following Saturday, when introducing other cells, look through the first and if any nucleus fails to have a virgin queen, it should be given another cell.

Number of Swarm Boxes Needed.

I have found one swarm box to the hundred nuclei sufficient. We have described the process up to the time of the emerging of the virgin queen, presuming we are using one swarm box. If the beekeeper grafts oftener than twice a week, more nuclei per swarm box can be run. If several hundred nuclei are maintained, it is advisable to have an extra swarm box to be used in "catching up" after a period of unfavorable weather, bad luck, mismanagement or accidents, which are liable to overtake the queen-breeder. I run six hundred nuclei and have eight swarm boxes. Very seldom are the eight all pressed into service. Usually but five are needed.

It is advisable to have always an abundance of cells, for, by the method described, they are easy to produce. To throw away fifty cells is much more economical than to be short five cells when needed. Not infrequently at the end of the season, I have a half bushel of cells that have been discarded. While it may look like waste to throw away hundreds of fine cells, it is the best way out. In the past I have sometimes tried to save them by using cages, etc.; but it is false economy. If the weather has been unfavorable for mating and you have a lot of ripe cells, while the virgins have not mated so as to be out of the way, just "junk" your cells and forget it. If the weather has delayed the mating three days you are set back just three days and you can't help it, unless you are a Joshua and can stop the sun while you catch up with queen orders; but most of us lack talent in that line. No, you cannot gain time by manipulated the clock, either. We have tried that. One of the ways in saving cells is to go through all the nuclei and cull out any queens that are not up to standard, giving them a cell; and should any virgin be missing, replace her with a cell also. In that manner part of the cells may be saved.

Another profitable way to use surplus cells is in requeening, as described in Chapter XXXI, where a double brood-chamber is used. If the single brood nest is in service, the old queen should be removed, and the cell given immediately. During a heavy honey flow the cell will be accepted without ceremony; but if no honey is coming in, a heavy feed puts the colony in condition to accept the cell perfectly. If one has extra hive bodies or nuclei, these may be brought into use and the cells saved. These hive bodies of nuclei may later be united with the colony to be requeened, after first hunting out and disposing of the old queen as described.

By carefully following the methods we employ for cell-building as we have tried to describe, fine large cells are produce at little cost, so that cells are very cheap, and a goodly supply for all purposes is constantly at hand.

Chapter XX. Misfortunes of the Queen-breeder.

From the grafting of a larva to the taking out of a fine laying queen is a long step. Many things may happen to that larva to prevent her "coronation." Even if the queen-breeder has done the very best he knows, there will be accidents and blunders. Things may be going along nicely when suddenly the weather turns cool, the rain pours down and the wind blows. Now, a little of this does not hurt much, but let it keep up for several days and it hurts a whole lot. Cells are torn down in finishing colonies and in nuclei. Virgins fail to mate and come up missing. Grafting time is on you and you are not expecting this "duck-hunting" weather to continue so you do not feed. The result is that when you graft, there is a very poor acceptance, and you go sulking around with the grouches, saying to yourself, "looks like I am just about out of the queen business."

Now if you get in a muss like this and you happen to be a commercial queen-breeder, you will begin to hear murmurings from the "office force." She will want to know why it is that, although the output of queens is falling off, the correspondence is rapidly increasing. She says, "Here are a lot of letters asking why you have not sent those queens as you promised; and here is one who says he is going to report you to the bee journals if you don't send those queens by return mail!" Well, it serves you right if he does, for not having a "tornado clause" in your contract; so you turn you back on the "office force" and tell her something like this, "Oh, give them some of that threadbare, moth-eaten dope about unfavorable weather conditions and so on!"

But the skies are clearest after a storm, so the clouds roll away and a soft warm south breeze comes up. In the morning you go out into the apiary, feeling that this world is not, after all, such a punk place in which to live. A meadow lark sits on a metal hive cover washing its feet in the heavy dew. As it points it's bill up to the sky and begins to sing, you wonder why it was a day ago you were in the dumps. Bees come scampering in with their baskets bulging with pollen. Occasionally they lose some of it at the entrance of the hive where it gets water-soaked and makes a mess in the doorway. The day is clear, the sun is shining, the bees are bringing in nectar, and queen-rearing goes on at high tension again.

Now what are we going to do when bad weather hits us? As our beloved James Whitcomb Riley says,

"It ain't no use to grumble and complain,
It's just as cheap and easy to rejoice;
When God sorts out the weather and sends rain,
Why, rain's my choice!"

Nothing we can do will make the conditions as favorable as they are during good honey-gathering weather; but we can do a few things that will materially lessen the loss. When it begins to rain or turn cold, a good policy is to feed the cell-finishing colonies as well as the colonies that are to furnish bees for the swarm boxes, and give a little to the colony containing your breeding queen. If the weather is very cold, contract the entrances considerably, and give the bees the feed rather warm. If the weather turns warm immediately the feeding was unnecessary, but no harm has been done.

Another incident that often "plays hob" with the queen-breeder's hopes is a queen-cell overlooked above the excluder, when a virgin has emerged and tears down all the cells in that hive. One cannot guard too carefully against this. Every few days examine the frames in the upper story of the finishing colony for queen-cells. Shake off all the bees from each frame and look closely. Sometimes a very small cell will escape notice if a careful search is not made, and the virgin emerging from it will play havoc with all the cells. Such cases as the above are when the extra swarm boxes must be brought into play in order to make up as soon as possible for the cells lost. But if care is taken there will be a natural surplus of cells, so that it takes an unusual setback to require more cells than you normally have ready.

Another and perhaps the greatest problem with which the queen-breeder has to contend is that of robber bees. When no nectar is coming in, these rascals make life miserable for the beekeeper and interfere with queen production seriously. If a nucleus becomes weak they rob it out and, stimulated by their victory, hunt all over the yard for others of insufficient strength to resist their attacks. When a hive is opened these robbers fairly swarm into it. They soon learn to follow the beekeeper around so as to be on hand instantly when he opens a hive. Even if a strong colony is left open too long, it is robbed out. Colonies from which bees are taken to fill swarm boxes will be robbed if not carefully protected; and when the queenless, broodless, method is used, the colony is powerless against them.

In our yard when robbers are exceptionally bad we remove the brood and queen from the cell-starting colony late in the evening and graft early the next morning. In that we "put one over" on the robbers.

Chapter XXI. Records for the Nucleus Hives

If one has even as small a number as one hundred hives, some system of keeping tab on the condition within is necessary. Some keep a book containing the record. I tried this, but after it got pretty well stuck up with propolis, it failed to function and was indeed a closed book. Keeping the record on the hive with a pencil is better. But too much time is taken to place the record there and too much time is required to read it, since it is necessary to get up close to the hive in order to see it. A record to be satisfactory must be so made as to be seen readily at a distance. I tried various schemes until the present "block system," which has given such entire satisfaction, was worked out.

Four conditions within the nucleus hive must be known. They are Queenless, Cell, Virgin and Laying Queen. To indicate these conditions we use a block of wood 2 ½ x 2 ½ x 1-¼ inches. It is painted white on top and red on the bottom; while the sides are painted black, red, white and blue, reading to the right. The block is laid on the hive with the white side up. When the nucleus is queenless, the black side is to the front; when a cell is given it is turned a one-quarter turn making it red; when the virgin emerges it is turned to white, and when she begins to lay another quarter turn brings it to blue. These colors may be seen at a glance from a distance and the condition known without close inspection. These colors are suggestive. Black for queenlessness, conditions are dark within. Red when the cell is given and danger for the cell that is being introduced. When the block is turned to white, the presence of a virgin is denoted, white being symbolic of purity. The laying queen is what you are working for and denotes a "blue ribbon".

Sometimes there may be other conditions in the hive that need attention, such as a lack of stores or bees.

When this is the case, the block is turned the red side up, which shows that it needs attention. The reading on the side of the black may remain the same. This record is very simple and effective, and has proved very satisfactory. Nearly every one has his own "system" in such matters.

Now we come to the matter of dates, which is equally simple but may not be quite so easy to explain. After a laying queen is taken out two weeks will pass before another can be taken out even if all goes well, which condition does *not* happen in every case. Consequently we divide the hive into four imaginary parts for in the two weeks' period there will be need of four operations-four introductions, four removals of laying queens, etc. As you stand looking at the side of the nucleus hive, the first position is at the left end, the second position is just left of the center of the hive, the third position is just to the right of center and the fourth position is at the right end. Let us suppose we are going to introduce cells on Tuesday. This is at the first of the week, so we put in the cell, place the block at the extreme left end of the nucleus and turn the red side toward us. The next time we are introducing cells we examine this, and if the cell has produced its queen we give the block a quarter turn to white, but leave it in the same position on the hive.

When the block is turned to white, it needs no more attention until the time comes to remove the laying queen. When we introduce cells the next Saturday we go to those not having cells, supply them and place the blocks just to the left of the center since it is the last introduction of the week. Then turn the blocks to red. Do this in a similar manner on the following Tuesday and Saturday, when the blocks will be placed at the left of center and at the extreme right of the nucleus hive for these days respectively. Now on the following Monday the queens that have emerged from the first cells we put in should be laying and should be removed and the blocks turned to black. On Tuesday we can give this nucleus another cell, or a cell may be given immediately upon removing the laying queen provided the feeder was filled with syrup or honey on the day previous.

If every cell we put in resulted in a laying queen, the position of these blocks would never be changed on the nuclei and they would only be turned to different colors. But if we are introducing on Saturday, let us say, and we examine one that is read on the previous Tuesday position and we find that the virgin is missing, we should have to move the block down to the Saturday position. All complications that might arise are taken care of with this system. Now suppose we come on Monday to remove a queen and she has laid only half a dozen eggs, not enough so that we can properly judge her merit. If we left the block in that position turned to white, we should be unable to tell whether she belonged to the lot that were ready to lay or the lot that were due to lay two weeks later. So in that case we turn the block to blue and leave it until the next shipping day and examine again. However, suppose the virgin when the shipping day comes is not laying, but looks as though she might in a short time, you would not turn it to blue for she is not laying, and if you left it at whit it would not be examined till two weeks later. In that case we turn the block one-eight of a turn, which brings it midway between virgin and laying, and thus it will be looked after when the next shipping date arrives. Now suppose we look for a virgin and cannot find her. Maybe she is out for a flight, as it is well known that, as soon as they are able to fly, they spend much of their time flying out for exercise before mating. You do not want to turn the block to white, for if the queen is lost, that nucleus will remain queenless for too long a period and you will lose the use of it until the shipping date arrives when you will see it is queenless. In such cases the block is given a one-eight turn backward which brings it half way between cell and virgin. The next time you are introducing cells you will examine it, and if you can still find no queen, a cell is given. If the queen is there, it is turned to white if a virgin, or blue if she is laying.

These blocks are made at the planning mill. They are of hard wood so they are heavy and will remain in place. At first, I had some misgivings as to whether they would stay in position. I wondered if the wind might blow them off. Soon after putting them there a terrific wind and rain storm came along that threatened to blow everything away. Not a single block was moved in the least. I never had a bit of trouble except once, and that lasted only a short time. I would find the blocks moved or turned around while some were on the ground. Was it possible the wind did do it after all? The children do not play up there, for discretion forbids. One morning the blocks were disarranged again. It had been a very still night with no wind. Now we have an owls nest in our oak trees every year, and that owl and I are not on the best of terms, for when he can catch me out in the apiary after dark, especially if I am bareheaded, he seems to take much pleasure in scratching me on the head. This owl is one of the smaller editions, about four inches in height, but golly! how he can scratch! I made up my mind it was that consarned owl, and now that I had a double grudge I had a good case against him and I could shoot him, so I prepared for Mr. Owl. But one day I happened to look out and I saw what the trouble really was. A blue jay was hopping from one block to another in the most jubilant spirits. Sometimes he would kick a block off, and sometimes he would spin one around. Now I do not want to shoot any of our birds that build their nests in our woods, but this blue jay was throwing the system all out of kilter. Why, he had some marked queenless when they had a laying queen, and he had one marked a laying queen when it had laying workers! So I had to ask myself this question, "Am I engaged in bee or blue jay culture?" Decidedly I was in the bee business, so I took my shotgun and as the guilty bird started to fly away I shot him. There were two nests of blue jays in the trees, but strange to say there never was another block disturbed. The owl continues to scratch my head heedless of the bricks and clubs that are shied his way; but since he in no way interferes with queen-rearing we get along pretty well.

Position of blocks on nuclei.

The first nucleus contains a virgin queen that emerged from a cell placed in it the first of the week. The second nucleus in the same hive shows the block in the queenless position. The queen has just been removed on Friday. The third nucleus illustrates the block turned to white, indicating a virgin that should be laying the first of next week. The fourth nucleus indicates that no queen could be found when looking for laying queens; but it was though the queen might be flying out, so the block is turned between cell and virgin. Study this picture and see if you can tell from the blocks what day of the week it is supposed to be and where we are now working. The position of the red and blue could not be shown. These hives are too close for mating queens, and are placed in this manner merely to illustrate the block system.

Colonies packed in quadruple cases furnish brood for forming nuclei early in the spring.

Chapter XXII. Care of Nuclei.

Too much emphasis can not be placed upon the importance of keeping the nuclei in the best possible condition. They must be strong in bees with abundance of brood and rich in honey if the best success is to be achieved in incubating the cells and mating the queens.

Honey in Nuclei.

If plenty of honey is provided in the nuclei there will be no absconding; but, if not, the bees will abscond about the time the queen should begin to lay. You should bear in mind that a nucleus in the proper condition is a valuable asset and will pay you big dividends; but if it is allowed to run out of brood, in it gets weak in bees or has not sufficient honey, every thing will go wrong. Cells will be torn down, virgins will be killed, and bees will abscond. Take good care of your nucleus. It is the goose that is laying the golden eggs. Do not starve it or it will cease to lay. Frequently frames in nuclei may be exchanged with advantage to both. In case one has an abundance of honey and no brood, and another has an abundance of bees and brood, these combs should be exchanged and both will be put in better condition.

Laying Workers in Nuclei.

If, for some reason, several cells that are given to a nucleus are torn down, all brood will have time to emerge and laying workers will result. To cure them of this bad habit, take away their comb and give them another with unsealed brood and all adhering bees. They will usually accept a queen-cell, and the trouble is over. However, as is the case in many other lines of human activity, prevention is better than cure. Laying workers never develop while there is brood in the hive. So, whenever you find a nucleus that has no brood some should be given the bees at once. This puts them back into normal condition, and they will accept a cell and no laying workers will develop.

To resume our program, let us consider that it is now Monday, the 28th of the month. If all has gone well, the queens reared from our first grafting will be ready to be taken out from the nuclei. The cells were introduced on the 15th. The queens emerged on the 16th, the virgins should mate on the 23rd or 24th and begin laying on the 25th or 26th. By the 28th they should be laying enough so you are able to judge as to whether or not they have all the appearances of a good queen, and yet they have not laid enough so that injury will result on account of their removal from the nuclei.

Laying Queens and How Injured.

These queens are now ready for shipment or to be introduced to colonies at any time you are ready. It is a well-established fact that, when a queen laying to full capacity is removed from the colony and placed in a mailing cage, she seldom makes good at the head of a colony as far as prolificness is concerned. When it is necessary to remove a queen in the height of her egg-laying, as for instance, if one wishes to ship a breeding queen that is in a strong colony, it is best to place her in a nucleus for a few days. In this case she can reduce the number of eggs she lays, gradually becomes much smaller in size and therefore she can stand the trip better. If the queen is extra large in size and laying to full capacity, I have found a splendid "reducing exercise" as follows: Take out two frames of brood with bees and the queen, and place them in a nucleus hive on a new stand. In two days' time, move the nucleus to a new location, and on the next day move it again. The bees that fly out do not return. The nucleus is thus weakened in bees and is not getting any honey or pollen; so the queen rapidly curtails egg-laying. To remove her when she is laying in a limited way does not injure the queen in the least, for this is in perfect harmony with bee nature. For instance in the case of a swarm, after the swarm has gone out, a few eggs will be found which the queen laid within a very short time before her departure from the hive. Consequently, if you can get the queen down to laying only a few eggs before removing her from the nucleus, she will not be injured in the least.

Chapter XXIII. Mailing Cages.

All features of queen-rearing are fascinating; but I enjoy looking for laying queens more than any other feature of queen-rearing. As some who read this are interested in commercial queen-rearing, and others who produce honey may at times find it advisable to ship queens to other yards, I shall describe the process of mailing queens.

For very short shipments the common three compartment Benton mailing cage will do. However, I believe even in short shipments the use of the large cage that has six compartments is advisable. This cage is good for any shipment in the United States and Canada. For export, I prefer a cage nearly square with nine compartments, each of the same diameter as in the regular Benton cage. These compartments should be about one half deeper however. The secret of good shipments is to have the cage of sufficient size so that it will accommodate plenty of nurse bees and still leave abundance of room for them. There should be space for every bee to have a footing on the bottom and top of the cage and still leave room enough so there can be a space all around the bee equal to it's own size. If the bees get too warm, they then all have plenty of room to fan and thus keep down the temperature. In the six-compartment cage, when the weather is hot, I find that about fifteen nurse bees give the best result. If the weather is cool double that number is satisfactory. If it is necessary to ship queens when the weather is cold or where the route is through a cold district, as for instance going over the mountains in early spring or fall, the use of the export cage is advisable. More bees can be put into the six-compartment size; but in case the bees pass through a warm district they would suffocate. The cage should be large enough so that the bees can spread out when it is hot, and yet large enough to hold sufficient bees to they may form a cluster in one or two compartments to keep warm in case the weather turns cold. Such a cage is ideal as far as size is concerned.

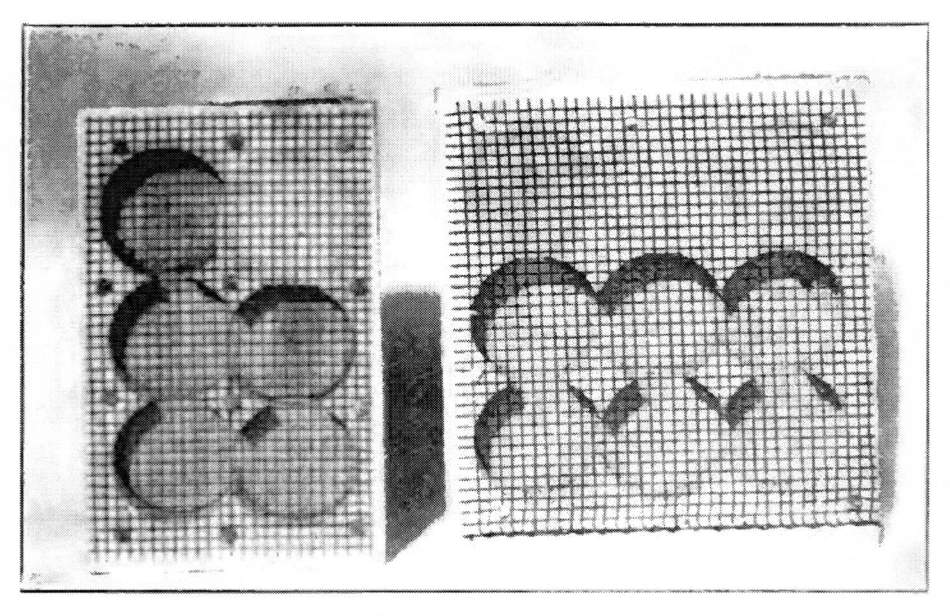

Queen cages.

Some years, the shippers of queens suffer a much heavier loss than in others. In the main, hot weather is the cause although the blame is usually laid on the queen candy. If the candy is made hard so the bees will not become daubed, it will give perfect results, at least for all ordinary shipments. Some have suggested that pure sugar candy lacks the nitrogenous food contained in pollen, and that this caused the loss in long shipments. Pollen could not be used in candy, as it would cause death by dysentery. Our friend, Allen Latham, queen-breeder of Norwichtown, Conn., has bee experimenting by mixing a little royal jelly with the candy as this would furnish the nitrogen and being predigested, would cause no intestinal disorder. We have been experimenting with this also, but have not gone far enough at this writing to come to any conclusion. However Mr. Latham's plan looks good.

We believe the principle cause of loss in hot weather is the lack of ventilation. Imagine a queen with her escorts, in a mailing cage with two little slits in the side of the cage packed in a mail sack, with all sorts of mail matter crowded all around the cage, and then perhaps with twenty-five or fifty more sacks piled on top of it and all out in the sun, with the thermometer reading one hundred in the shade. The wonder is that any survive. During very hot weather we use and recommend what we call "the hot weather cage." It is the regular mailing cage with a heavy fibre cover raised about one-eight of an inch with wooden cleats. This remedies much of the hot-weather troubles.

Hot weather cages.

Chapter XXIV. Queen Candy.

The other important feature in the perfect shipment of queens is the food during transit. Much trouble is experienced in shipping queens with candy not sufficiently hard. I have experimented quite extensively along this line by keeping a number of cages containing worker bees and different kinds of food in the attic, the basement and various other places under all sorts of conditions and temperatures, and found the chances of success greatly in favor of the harder candies. Not realizing how decidedly candy affects the bees, many make the mistake of using candy too soft. Watch the bees in the mailing cage. You will see them continually rubbing their sides against the candy. If it is a trifle too soft they gradually get it on their sides and into their breathing tubes. Then they fret, which makes a bad matter worse, and they die from worry and suffocation. Upon examination the dead bees do not appear candy-daubed at all, but that is what really caused death.

The commonest method of making cage candy is to take honey or invert sugar, put it in a vessel, set it on the stove and heat to a temperature of about 140 degrees. Then stir in as much powdered sugar as possible, put it on the bread board and knead in more powdered sugar until very, very stiff. You cannot get in too much sugar or knead it too thoroughly. Another difficulty is encountered in the use of this sort of candy made with powdered sugar. As is commonly known, powdered sugar contains from three to five per cent starch to keep it from becoming lumpy and hard. This does not make the best grade of candy, for starch causes dysentery to bees in long shipments, but for short shipments no harm is done. Powdered sugar without starch may be obtained by a special order, but it becomes so lumpy after a few days that fine, smooth candy cannot be made with it.

However, good candy can be made with either the sugar containing this small amount of starch or the pure powdered sugar. The pure sugar is to be preferred. During the war when powdered sugar could not be produced we made a fondant out of granulated sugar that gave excellent results; but it is a very exacting process to make it, so that we abandoned it. The candy that we consider the very best, however, is made from pure powdered sugar and homemade invert sugar.

As to the use of honey for cage candy, it is generally acknowledged that American foul brood has been spread in this manner. Even if the present laws are complied with, there is still danger. If for instance the shipment is accompanied with a certificate of health from an apiary, the candy may have been made of honey from some other apiary, disease infected; or, the diseased honey may have been produced some time before when the apiary was infected. The source and spread of bee disease are so mysterious and disastrous that we cannot be too careful of contagion, and while the party using such diseased honey would of course be ignorant of the fact, the damage is done nevertheless. Therefore not only should the breeder be particular but the purchaser should also exercise precaution and, for introducing the queen, he should never use the cage in which she is shipped. Let him always transfer the queen to a cage of his own and burn the shipping cage.

As to boiled honey, it is about the worst thing which can be used in making candy. Bees that have been fed on it for only one day frequently show distended abdomens, indicating they will soon die of dysentery.

For ordinary shipments, invert sugar has been used instead of honey. This may be purchased from dealers in beekeepers' supplies. With the process of making the very best grade of queen cage candy without the use of honey, there is no reason why honey should ever be used in making queen candy at all, even for extremely difficult shipments as is the case in export trade or in very hot dry weather. It has been considered that a fine quality of white honey is superior. However, the chemist tells us that honey and invert sugar are the same as far as the process of drying out is concerned. Now we know that commercial invert sugar is not so thick as honey, and that it does dry out more than honey. Pure water boils at a temperature of 212 degrees. If a heavier substance such as sugar is added, it requires a higher temperature to bring it to a boil. By this test one can easily prove that the commercial invert sugar contains more water than honey, which causes the candy made with it to dry out sooner than honey. The commercial invert sugar boils at a temperature of 245. After this fact was learned some invert sugar was made by using granulated sugar and tartaric acid, and it was boiled till it reached a temperature of 250 degrees, which is five degrees above the boiling point of honey. This gave a very thick, heavy syrup, so thick that it had to be warmed before it could be handled, which induced me to do some investigation along this line. The result of these experiments led me to make my own invert sugar, which is thicker than honey and therefore is superior to the best honey for queen candy. As was expected, candy made with this and powdered sugar does not dry out at all, no matter how hot and dry the weather is.

Receipt for Making Invert Sugar.

Put one-half pound of water in a kettle and place over the fire until it comes to a boil. Then add one pound of granulated sugar, either beet or cane sugar, they being exactly alike. To this add about ten grains of tartaric acid. As few will have pharmacists' scales to weight the acid, you can use about one-eighth of a teaspoonful or a small amount on the tip of a table knife. If trouble is experienced in getting the right amount, an empty 22-short rifle cartridge may be used as a measure. One of these holds 2 ½ grains, therefore four of these level-full will furnish the right amount. A little more or less will do no harm. Put in the acid and allow the syrup to boil slowly without stirring until a temperature of 250° is reached. Then pour it into a Mason jar. The cap should be kept on the jar as this syrup, being so thick, will absorb moisture if left exposed to the air.

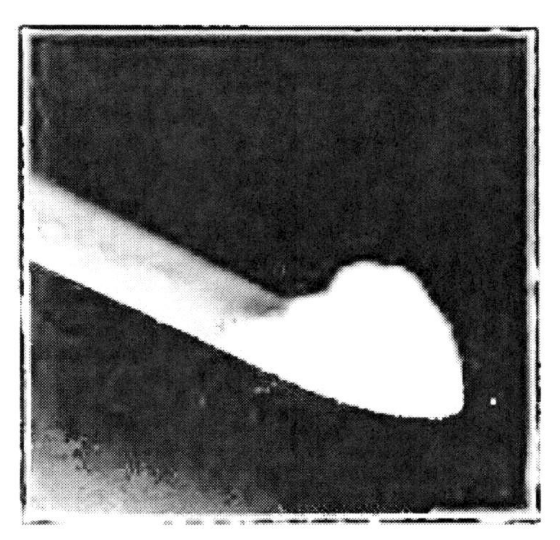

A small amount on the tip of a table knife.

Making the Candy.

Take the bread board and cover it with powdered sugar two or three inches deep. Warm the invert sugar slightly, not above 120°, so the dough will work up easily. Pour a little invert syrup on the sugar and cover it with more sugar. Then work the dough up into a ball and knead the dough until it is so stiff one can scarcely pinch off a bit between the thumb and the finger. In fact, I have never been able to make this candy too stiff. Considerable kneading is required to get it just right. If the air is moist the candy becomes sticky, and more sugar should be worked into it before provisioning the cages. This candy will keep indefinitely. One winter we placed some above the hot-air register of the furnace, and in the spring it was still soft and the bees ate it readily. If worked long enough it will become quite dry, more like bread, but will not become hard. In that condition it will not gather moisture and will not run. As pure powdered sugar will become lumpy, if kept on hand too long, it is advisable to order twenty-five or fifty pounds and make it all up at once. When first made, if the weather is damp, it is well to keep the top covered with a little dry powdered sugar.

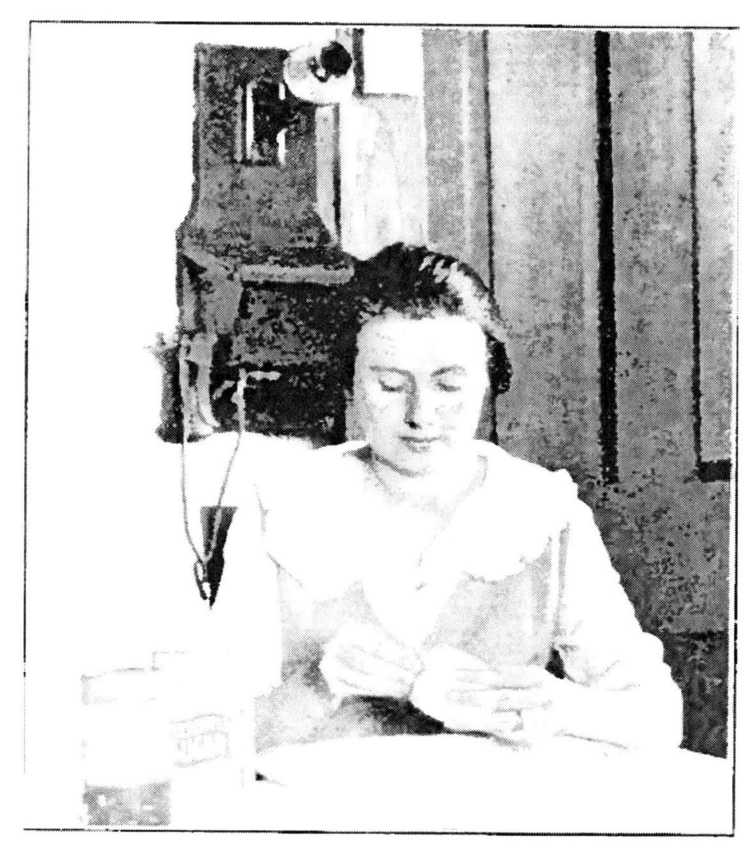

Pinch off a bit between the thumb and finger.

This makes it very convenient to pick up the nurse bees.

It is quite unnecessary to say anything about putting the queen candy into the compartments other than some suggestion as to the amount required. For shipments anywhere in the United States or Canada, one compartment full of candy is plenty if twenty-five or thirty bees are put into the cage. If fifty or more are placed in the cage or if the shipment is two thousand miles or over, two compartments should be filled. For export, it is well to fill three of the compartments in the export cage.

After provisional, the perforated tine that covers the hole in the cage should be swung around the tack placed in it so that it may be easily removed with the fingers. Having the cages provisioned, place them in the hive-seat on the left side. Open the nucleus hive and give the bees a very light puff of smoke from the smoker. Lift out the frame and find the queen. Pick her up with the right hand by taking the two wings between the finger and the thumb. With the left hand put the brood-frame a short distance down into the nucleus hive and tilt it over away from you so it will remain there. This makes it very convenient to pick up the nurse bees. Take up the mailing cage with the left hand and put the queen into the cage, closing the hole with the forefinger of the left hand. Then proceed to pick up worker bees by the wings and poke them into the cage, head first, closing the hole in the cage after each bee with the finger as before. When one gets accustomed to it, he can fill a cage in half a minute. Once in a while the beekeeper will get hold of a balky bee that insists on bracing with its front feed and refuses to go in. When the pressure is brought to bear with the finger, the bee runs out its sting, with the result that in time that finger becomes somewhat callous. A thimble might be used on the left forefinger, but this is awkward and unsportsmanlike.

The Question of Cataleptic Queens.

In handling queens, many beekeepers have observed that once in a great while a queen suddenly becomes unconscious and sometimes dies. The reason assigned is that she took a cramp or had a cataleptic fit.

The author has observed this for the past twenty years. Some seasons it would occur but once; during others half a dozen times or more. In my mind there has always been some doubts whether or not the queen was cataleptic. One season, the loss was heavier than usual. One day two were lost. I say "lost" as the queens were always discarded after having a "fit," for previous experience had made me believe that they were permanently injured by having these "spells."

The day the two queens were lost, I observed how very similar was the action of the injured queen to the one that had been stung by another queen. There was a sudden collapse, then a slight quivering of the legs. In one case this lasted for over half an hour, when the queen slowly revived. In the other case, the queen quivered four about the same length of time and then died. It seemed certain to me that in some mysterious manner these queens were getting poison from a sting. Could it be that the poison on my fingers from worker-stings was causing the mischief? Investigation failed to substantiate this. I noticed that in one case the queen had taken hold of the top of her abdomen with a front foot, which might indicate that she had received a slight prick in the foot from her own sting. I therefore watched carefully and soon this belief was confirmed. The queen in taking hold of the tip of her abdomen exposed the sting. Then, in trying to get hold with the rest of her feet, she would strike right at the point of the sting. In this manner she undoubtedly received some of the poison. Since that time we have taken great care that a queen is not allowed to take hold of the tip of her abdomen, consequently no more queens have been afflicted with fainting spells.

Chapter XXVI. Clipping Queens' Wings.

There is considerable controversy on as to whether it injures a queen to clip her wings. This controversy has been on ever since the practice of clipping was started. Some claim it injures the queen, and some as stoutly maintain it does not. My experience leads me to agree with both factions. The following article, written by me and published in "*Gleanings*," tells of my first attempt at clipping a queen.

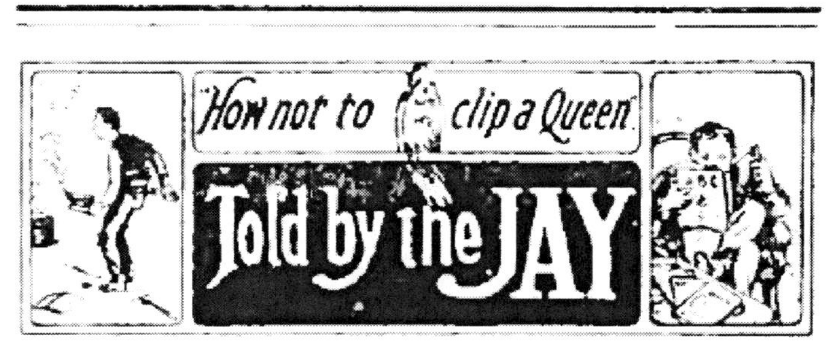

AN AMUSING AND YET NOT UNCOMMON EXPERIENCE OF BEGINNERS.

The first colony of bees I got was in the ten-frame homemade hive, which I kept standing in the back yard the first year, not daring to go near it. In the fall, I went out one night and peeped under the cover, and was surprised to see that there was no honey. I supposed all a fellow had to do to get honey was get some bees and they would do the rest. Nothing succeeds like success, so they say. Not so here. Nothing makes me succeed like a failure, so I determined that next year those bees should make some honey or furnish a reasonable excuse.

I subscribed for Gleanings and got the ABC. Then the bee fever took hold of me in earnest. I studied the book night and day. I knew it all by heart. I got the "Facts About Bees," and learned it till I could recite it as easily as a minister can quote scripture. The argument in it was good. Everything in it was all worked out. How grateful I felt that everything had been learned for me, and all I had to do was to reap the benefits! I tried hard to be unassuming but inwardly I could not help feeling proud that I knew everything about bees.

Now the book said: "Catch the queen and clip her."

I did all this studying in the winter; and how I longed for spring to come that I might demonstrate what I already knew! How that winter persisted in staying with us, and how reluctantly did spring show her shining face! But at last, in the latter part of March, there came a beautiful, bright warm day-just the time for clipping the queen! I had never seen a queen, and my anxiety to view her majesty was something fierce. I had an assistant cover me with mosquito bar. I put on mittens and wrapped my wrists with rags. Then I fired up the smoker and prepared to go into action.

How I dreaded opening that hive! I felt a little pale, but my teeth were set and it was do or die. I was too big a coward to retreat while every one was watching. I must have been an aw-inspiring sight to those bees as I swooped down upon them dressed in armor, with the smoker spitting smoke and fire. I soon enveloped the hive in smoke, gave it a few jolts tore off the cover, then smoked again. Of course the bees cowed before such a vicious onslaught. Now, the books said, "Catch the queen and clip her." Clipping was the primary object of the expedition; but I saw where the books were right in saying "Catch the queen" before saying "and clip her." The only change in the wording of that I would make would be to precede that with "find the queen." I took out the frames carefully, and stood them around the hive in various places; but could not "catch the queen." I looked and looked. There were more bees in that hive than I had expected to see in ten hives. The separation of a mixture of the proverbial haystack and the needle would have been a cinch compared with the task in hand. I hunted all the afternoon, and had to give up on account of darkness. I was disgusted but not discouraged. This problem confronted me, "If I fail to find one queen in half a day, how long will it take to find several thousand queens?" (the number I expected to have in a year or two).

Nothing succeeds like a failure, and the next day I went after them with more zeal than ever. On lifting out the third frame my eyes rested on a bee the like of which I had never seen before. It was a long bee, and she walked with a more majestic treat over the comb and did not seem to be in such a rush as the rest of the bees. She was of a dark-brown color, and how handsome she looked! Verily this must be the queen! The queen of Sheba might have looked good to Solomon; but she was not arrayed like this one.

The next thing, "Catch the queen." I tried to make the catch, but she was not so easy. Just as I would close my fingers on her she was not there. At last I got hold of one wing but she buzzed around so that I let her drop.

If about one third of the wings is clipped off no harm can possibly come of it.

Again I got her by the wings and tried to transfer to the left hand, but her head did not stick out far enough for me to get a good hold, and she backed out and got away. Next time I shut down so hard that I was afraid I would kill her, and then let up so that she got away again. This time she dropped in the grass and I had a time to find her. The fourth time I held her between my left thumb and finger in a trembly fashion, much as a dog bites a rat, and probably the sensations to the rat and queen were similar.

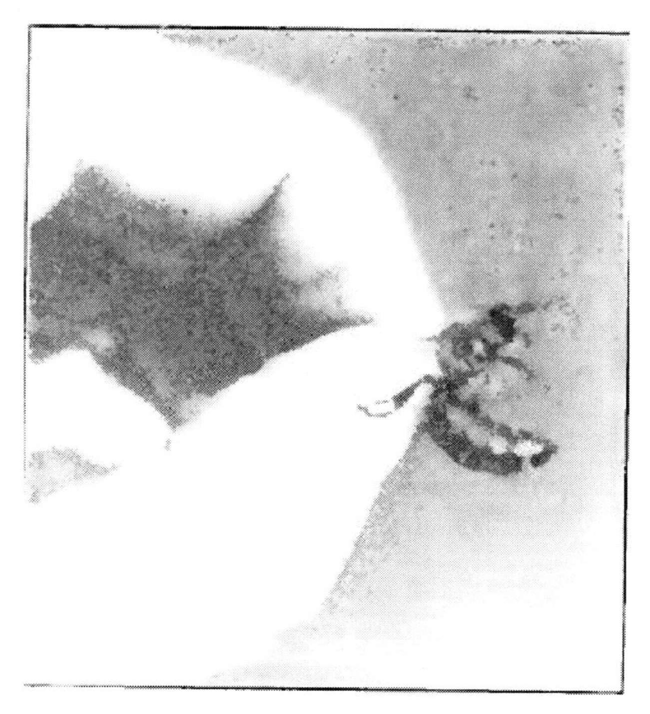

Pick her up by the wings.

I then got the shears. I forget whether they were a large pair of tailors' shears or the kind they use for shearing sheep. In my enthusiasm I had used them in prying frames apart, and they were more or less gummed up with propolis. I slid them under the wings and shut down. The wings bend over, but would not cut. I tried again and again until I either wore the wings apart or pulled them off. But I got them off and a leg with them. A little later I thought I would "shook" the bees into a new Danzenbaker hive, and I was astonished to find that they had a new yellow queen with wings of the regulation length and a full quota of legs.

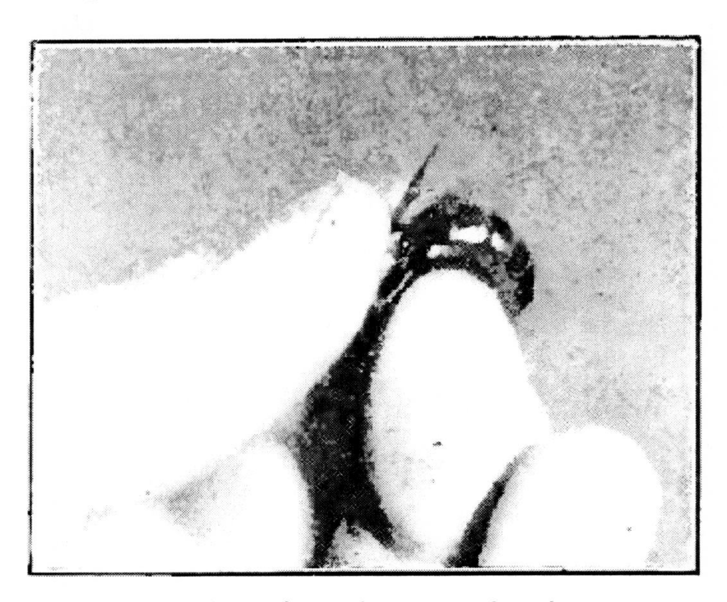

Leaving the wings projecting.

In the case described it undoubtedly was injurious to the queen to have her wings "cropped" as one correspondent puts it. However, if care is taken and clipping done properly it can be no more injurious to a queen to have her wings clipped than it would be in case the wing of a hen is clipped to keep her from flying over the fence and getting into the lettuce bed-it all depends on how it is done. If the wings are clipped too short, injury may result. In such case nerves and arteries may be severed. If about one-third of the wings is clipped off, no harm can possibly come of it. This bears out in practice, for in thousands of queens clipped in this manner, I never knew of one injured. The bees treat such queens in exactly the same way as though they had their entire wings and do not supersede them any sooner than the others.

Clip off about one-third.

In clipping queens' wings some prefer to hold her majesty by the legs, while some prefer to hold her by the thorax between the fingers and the thumb of the left hand. After trying both I prefer the latter method. To clip a queen, first pick her up by the wings with the forefinger and thumb of the right hand. Then pass her to the left hand, placing the thumb on the upper side of the thorax and the forefinger on the lower side. In this position her abdomen forms a curve over the end of the finger, leaving the wings projecting in such a manner that they may easily clipped. Then pass a blade of the scissors under the wings and clip off about one-third of the length. If preferred the wings on one side only may be clipped. It is not necessary to clip off much of the wings in order to prevent the queen from flying, as a laying queen has all she can do to fly when she has all of her wings in good order. The queen must never be clipped before she begins to lay, for of course, since she mates with the drone on the wing, she could not mate if her wings are clipped. As she never mates after she begins to lay, no harm is done by clipping her wings as described after eggs are laid. The wings of the queen do not grow out again after clipping, so one operation is sufficient for her lifetime.

Lest the beginner might not thoroughly understand the significance of the above, let it be understood that a queen must under no circumstance be clipped before she begins to lay, for it cannot be definitely ascertained that she has mated and fertilized until she begins to lay. As the queen mates only upon the wing, it can be understood that if her wings were clipped she could not fly and therefore could not be fertilized. There are frequent cases where the amateur catches a swarm, and, if the queen is found, she is clipped to keep the swarm from absconding. One should be careful in such cases to be sure that the queen he clips is a laying queen, for, if the swarm should be an after-swarm with a virgin queen, clipping would render her incapable of mating. In such a case if the swarm were not given another queen it would be lost. In some cases the virgin would become a drone-layer and in others she would be killed by the bees who seem to think it was the queen's fault in not going out and mating. After the queen is killed by the bees, the colony soon runs to laying workers and is worthless.

Chapter XXVII. Introducing Queens.

Nearly thirty-five years ago, Mr. G.M. Doolittle wrote, "Perhaps there is no one subject connected with beekeeping that has received so much notice in our bee papers and elsewhere as has the introduction of queens." We find the condition in this respect very much the same today. Methods come and flourish for a time, and then quietly vanish. Many of these so called "new methods" were used and discarded before any of us were born. I have been guilty of making some startling discoveries only to find that they had been known many years ago and discarded because they were of no account. Other interesting facts I have come upon, only to realize after a certain period of time that they were advocated by others years ago. For instance, I proved to my own satisfaction that the presence of queen cells in a hive has little or no effect upon the bees' accepting a queen, provided no virgins emerged before the queen was laying. This is contrary to the general belief and I thought I stood alone in this, yet Mr. Doolittle says the same thing in "Scientific Queen Rearing." Lately the "Honey Method" of introducing queens has come forward and was thought by many to be something new, but Mr. Doolittle also describes that. Yes, and it is just as unreliable today as it was forty years ago. One year I used tobacco stems in my smoker to introduce queens, closing the entrance until the bees came out of their debauch. I thought I had something good as well as new, but we find that Alley used it many years ago. I think it is the best of any "smoke methods," but it is poor enough at that. We can all avoid traveling in a circle in this manner if we will only read all of the older publications in beekeeping. The old "masters" can teach us much. Many times we can find the very thing we are getting ready to "invent" and also the reason why it is "no good." All kinds of torture have been inflicted upon the innocent queen and bees during queen introduction. There are the "Starvation Method," the "Drowning Method," the "Honey Daub Method," "Peppermint," numerous "Smoke methods," "Chloroform," "Carbolic acid," etc. Yes, and away back in 1744 the English had the "Puff Ball Method." When one of these balls was "puffed" at the bees they became unconscious. Dr. Phillips tells me that they claimed it made the bees "forget," and when they came out of it they had forgotten all about the queen question and couldn't remember whether the queen they then possessed was the one they always had or not. No doubt many of the queens introduced in that manner also forgot to lay, and many would "forget" to live.

All of the above "heroic" methods remind me of the way the students of Anthropology tell us the cave man who lived a few hundred thousand years ago got his wife. Now the cave man was not at all sentimental in his make-up but rather intensely practical and conducted his business affairs in a drastic manner. He did not believe in all this serenading by moonlight, neither did he ever buy 30c gasoline to take his prospective bride joy-riding in his flivver, and he had no use for this swinging-on-the-gate stuff. Not he! When he wanted another wife to add to his heterogeneous collection, or if one was getting old and he thought she needed superseding, he picked up his club and went after another wife. As the people in those days were vegetarians, he knew where to look for one; so about supper time he would find her out in the alfalfa or sweet clover patch enjoying the evening meal. He walked up behind her, swatted her over the head with is club, and carried her back to his cave where she became his dutiful and obedient wife.

Now, he got a wife, sure enough and maybe it is a matter of sentiment, but don't you know I like the way we do it nowadays much the better! So likewise I like more sentimental methods of introducing queens. If the heroic methods just mentioned were sure in their results, we might overlook the rough treatment; but they are not. There is no method that will permit the taking away of a queen and immediately releasing another in the colony with anything like certainty, for it is entirely against bee nature and you cannot change bee nature. Let us try to understand and work in harmony with it. I am sure better results can be obtained.

Loss in Introducing with the Mailing Cages.

Probably the method used today in the majority of cases is that of introducing the queen in the same cage in which she was shipped through the mail. The loss of queens by this method has been frightful. Man who have had many years' experience as inspectors and are in position to know, have told me that they believe 50 per cent of queens are lost in this manner. An expert honey producer in California told me that he had kept track of his loss and found it to be one out of three in the introduction of all queens he brought through the mail. Various reports have come to me where six queens out of twelve are lost by using the common mailing cage as an introducing cage.

**(My friend M.T. Prichard, one of the best queen-breeders in the country, says that mailing cage is quite reliable for introducing provided the queen is kept cages 48 hours before the bees get at the candy to release her. He says if this precaution is observed the loss will be negligible. He recommends plugging the candy hole with a plug of beeswax for two, and in some hard cases, four days, at the end of which time the plug is removed and fresh candy put in.) Now if the bees would kill the queens outright the damage would not be so great, but frequently the queens are injured so they never make good; yet they remain at the head of the colony, possibly a year or more,* *losing for the beekeeper the surplus that colony would have made had the queen been properly accepted.*

It has often been a wonder to me how the beekeepers have been content with this heavy loss. If a stockman in buying cattle lost fifty per cent or even five per cent by having the cattle fight when he united the herd, he would look for better methods. The loss to the beekeeper is not so much in one lump, but the percentage is the same whether it is cattle or bees. In this connection we are reminded of the story told of a man who, in telling his friends how to teach their boys to swim, said: "All that is necessary to teach a boy to swim is just to catch the kid and pitch him right into the deep water. He will swim all right, for he has to. I know that method works, for I taught my eight boys to swim in just that very way, and lost only one out of eight!"

However, if this method of queen introduction must be used, the best way, in our opinion, to use this cage is to remove the queen from the colony to be requeened, take out a frame and set it away in the honey-house spread the brood frames apart, place the cage between the frames with the wire screen downward, and press the combs tightly against the cage to hold it in position. In this manner the bees cluster on the wire and get acquainted with the queen. It is well to tack a piece of tin over the candy for a day so the bees cannot release her too soon. When there is a honey flow on, this method will be successful, probably in four cases out of five. However, I believe this is the most unreliable method of any that I shall describe.

The Doolittle Cage.

The Doolittle cage is better. To make this, saw off two pieces from a broom handle, one five or six inches long and the other piece one inch long for the ends of the cage. Then construct a cylinder of wife-screen cloth the size of a brood handle to fit between the ends. The wire screen is tacked permanently to the short piece. A hole about three-eights of an inch in diameter, to be filled with candy for introducing, is bored through the long piece, which is withdrawn from the wire cylinder in order to put the queen into the cage. To introduce a queen she is first transferred from the mailing cage to this one. A frame of honey is taken out to make room and the cage is placed down between the brood-frames, which are pressed together to hold it in place.

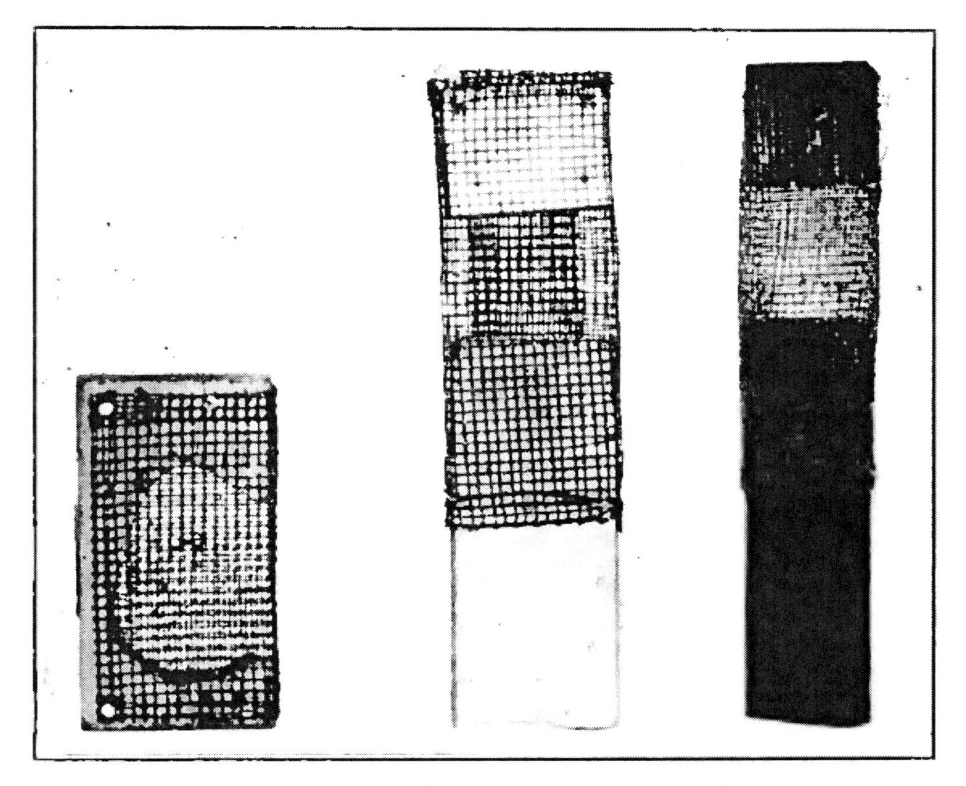

The Doolittle cage and two forms of the Miller cage.

The Miller Cage.

Dr. C.C. Miller constructed a cage similar in principle to the Doolittle; but he made his flat, using two wooden cleats so that he could shove it into the entrance or place it between frames without removing a comb to make room for it. A modification of this cage is shown with the queen excluding attachment. Either of these gives some better results than the mailing cage, but still there is considerable loss when conditions are not just right. Many of the Doolittle and the Miller cages are in use throughout the country. Some prefer one, some the other. With their use, the mailing cage is discarded, thus preventing the possible spread of American foul brood, as this disease has been scattered far and wide through diseased honey in the mailing cage.

The Push-in-the-Comb Cage.

Realizing the shortcomings of the above methods and reviewing the methods used in introducing queens, there seemed to be one style of cage that gave nearly perfect results, and that is the cage know as the "Push-in-the-comb cage." Mr. Doolittle used this cage and stated that not one queen in a hundred was balled when it was used. To make this cage take a piece of wire-screen cloth about six inches square and cut a notch out of each corner, so that when the edges are bent down it will form a bee-tight cage with one side open. The queen is placed on the comb, the cage is put over her and pushed down into the comb. In two days the bees usually burro under the cage and release the queen. If she is not out by that time, the cage is taken out and the queen released. This cage gives much better results than any of the others described, but it has some undesirable features. Sometimes it comes out of the comb or the bees gnaw away the comb, release the queen prematurely, ball and kill her. Moreover, since there is no convenient way of getting the queen into the cage, some are lost.

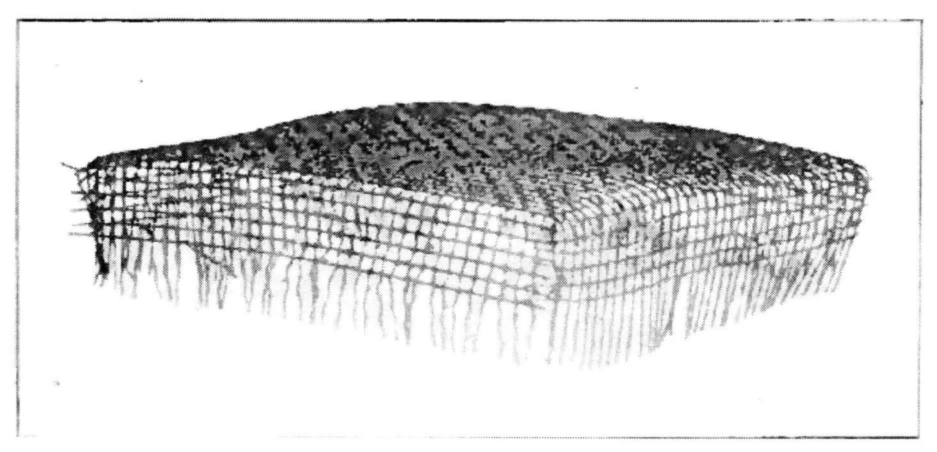

Known as the Push-in-the-comb cage.

Smith Introducing Cage.

Our Introducing Cage.

Realizing the heavy loss in queens, I have been experimenting for a number of years to perfect a cage that is sure in its working by using the principle of the Push-in-the-comb cage, and overcoming the objectionable features. I believe I have succeeded.

Our cage is made of half-inch material in a rectangular form. On the inside of the frame is nailed a strip of heavy tin so cut as to form pointed teeth. Wire screen is tacked on the top. Two holes are bored through the frame, one in the end and the other in the side. Corks fit snugly into these holes. On the inside of one hole is a piece of queen-excluding zinc, while the other hole is used to admit the queen.

And press it in tightly.

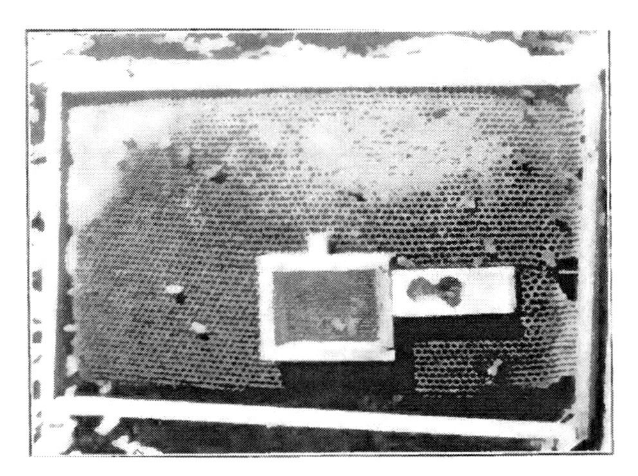

Now take the cage containing the queen and allow her to run up into the introducing cage.

To introduce a new queen go to the colony to be re-queened and remove the queen. Select an old comb that has had many brood reared in it. In such a comb the midrib is tough, for it is covered with old cocoons. If the colony to be requeened does not contain such a comb, get one from another hive. This old comb is important. Stick the cage into the comb, place it between the knees and press it in tightly. Now take the cage containing the queen and allow her to run up into the introducing cage. See that both corks are in place and set the comb back into the hive. It will be necessary to remove a comb of honey to make room for the cage. In two remove the excluder cork and allow the bees to get to the queen, and in two days more the cage may be removed and the queen liberated and all is well. In extreme cases where the robbers are bad and one had exceptionally cross hybrids or black bees, it is best to leave the cage in the hive for three days before removing the cork over the excluder and for three days more after that, but I have never found this necessary. Space should be left between the cage and the comb next to it so the bees can crawl over the wire screen and get acquainted with the queen. If queens that are to b introduced have been received through the mail, it is a good plan to burn the cages unless the candy in the cages contains no honey. American foul brood has been scattered all over the country in queen-cage candy.

As soon as the cork is removed over the excluder the queen increases egg-laying at a rapid rate, because the bees can get to her and feed her. She lays in all the empty cells, then goes back and lays in them again and again, sometimes almost filling them with eggs. She, therefore, is laying at full speed, and when released from the cage will fill the frames with eggs at an astonishing rate. I have frequently found eggs in three frames the day after the queen was released.

The queen lays more eggs the first day after being released than she would in two or three days where she could lay no eggs in the cage, and had to build up to egg-laying after being released. She has greatly increased in size, so by the time she is turned out she is a large laying queen, with the odor of the colony; the bees have been feeding her and have passed in and out of the cage so that the queen is as much the mother of that colony as though they had reared her themselves. When this cage is used as above described, I have yet to lose a single queen. Others have done equally well. In fact, I do not believe a queen would ever be killed by the bees when the cage is properly used.

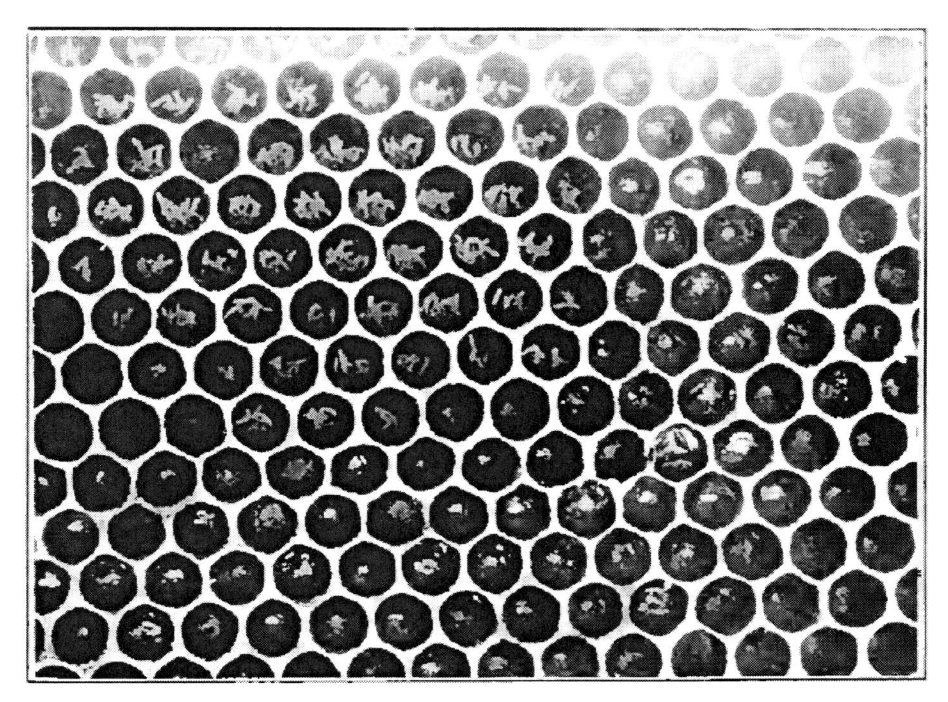

And lays in them again and again.

Upon several occasions, I have had queens killed because I had overlooked some cell from which a queen emerged and killed the one that I was trying to introduce. Upon several occasions, virgins from nuclei got into the hive, were accepted, and killed the laying queen as soon as she came out of the cage. Once when I was making increase I shook the bees into a new hive at the time I released the queen. She came up missing. There were robbers who probably killed her. I have had the same thing happen to a queen that had been in the colony for a year, so I do not consider that the losses mentioned were in any way the fault of the method of introduction. To avoid the possibility of having a virgin in the colony, one should examine the cage at the time of releasing the queen, and if there is another queen in the hive the bees will be balling the cage. In such cases hunt up the other queen and remove her, leave the cage in a couple of days more and the queen will be accepted. I have saved several queens in this manner.

Reasons for Acceptance.

This is the "sentimental" method, and it in no way injures the queen. It is well known that an old and failing queen can be introduced to almost any colony as the bees pay no attention to her. From the fact that the bees know they have a queen while she is in this cage and that she can not get out and lay in the combs in a natural manner, I believe they consider they have a queen that needs superseding, which is another reason they accept her so readily. I have noticed that at times they build a piece of comb in the space left vacant by the removal of the frame, and on this comb they start numerous queen-cells, expecting the queen to lay in them. Sometimes a dozen or more cells will be started on a piece of comb not more than four inches long and two or three inches wide.

Introducing Queens to Laying Workers.

As a rule, it is not worth while to try to introduce a queen to laying workers from the fact that they will not readily accept the queen and, even if it is successful, these old bees are not capable of acting as nurses; so that if it does build up at all, the colony is very slow in doing so. However, the following method had been entirely successful in putting the laying workers back in the job in a satisfactory manner: Take a frame of emerging brood from another colony and use the cage on that as previously described. Set this in the center of the colony of the laying workers and introduce as before. The bees readily accept the queen, and the frame of emerging brood furnishes nurse bees enough to give them a start. If late in the season and the colony is weak, it is well to give it two or three frames of brood at the time of introducing the queen.

Emerging (or Hatching) Brood Method.

One method without the use of any of the above-described cages has been used for many years with almost perfect success. It is that of placing the queen on combs of emerging brood after having brushed off all the bees. The main objection to this method is the time and work it takes. However, if one has a very valuable queen and does not mind the work, it gives excellent results.

The procedure is as follows: Take four or five frames of brood and put them over a strong colony above a queen-excluder. In ten days all of the brood will be capped over. Now take off these frames of brood from the colony and be careful to brush off every bee. Place them in an empty hive and stop up the entrance with rags so that no bee can get out. Take this into the house, remove the perforated tin on the mailing cage containing the queen, set the cage down on the bottom-board and close the hive. The queen and bees will crawl out of the cage on to the combs. Keep this hive in the house so that the temperature is even in order that the bees may emerge and not become chilled or overheated. In five or six days set this hive on a stand, open the entrance just wide enough for one bee to get in and out at a time. Watch to see that robbers do not overpower it, as it will be a couple of weeks before the bees can put up much of a defensive fight. Do not allow your curiosity to get the better of your judgment and induce you to open the hive for several days after you put in the queen, for, if there are not many bees emerged, the queen is apt to fly out. I once lost a fine imported queen in that way. This method gives excellent results when carried out as above described.

Emerging brood may be taken directly from the hives instead of placing them above a queen-excluder, but in that case the unsealed brood crawl out and die and make a mess.

Unsatisfactory Modification.

Some have suggested a change in the method and recommend that, instead of putting the hive with brood into the house, it be set over a wire screen above a strong colony in order that the bees may get the heat from the colony below. In theory this is fine, but in practice it is a pronounced failure. Many of the queens will be found dead when this method is followed. It does not look reasonable that the bees sting the queen through the wire, but that may be the cause of the death of the queens. My assistants in nailing up queen cages frequently get their fingers stung through the wire screen. One day when the weather was rather cold, I had a mailing cage containing queen and bees. I placed my hand over the wire screen to see if it would warm them up. Immediately a bee planted it's sting in the center of my hand leaving it there. It is possible that the bees sting the queen through the screen. At any rate many have reported losing queens when introducing them in this manner, above a strong colony over a wire screen; but the former method of removing brood to the house is an excellent one.

A Common Cause of Failure in Queen Introduction.

Regardless of what method is employed in introducing queens, it is very essential that all combs be put back in the same position that they were before removing. If this is not done, many queens will be killed even after they have been accepted and have been laying. When the combs are not put back as they were originally found, this is what is apt to happen: When the new queen is released from the cage she takes a "swing around the circle" to see to it that all queen-cells that have been started are destroyed. Now, if a comb of honey has been inserted in the middle of the brood-nest, the queen does not realize that there is more brood over the other side of this comb; consequently she does not go over to that side to attend to the destruction of any cells that may have been started there. She seems to think that she has been all over the brood-nest, and settles down to egg-laying. A few days later a queen emerges on the other side of the comb of honey, and sooner or later the virgin and the laying queen meet and the laying queen is always killed. I lost many queens in this way before I discovered the cause. Many who buy queens through the mail have trouble from this cause. They will wonder why it is that although the queen was laying profusely, she was "superseded" so soon, for she was found in front of the hive dead and a virgin was discovered in the hive. She was not superseded; she was killed by the virgin that came from the other side of the hive.

Why Queens Die in the Mailing Cage.

When the common mailing cage is used as an introducing cage, in some instances all the workers and the queen are found dead after the cage has been in the hive. We have had many inquiries as to the cause of this loss. I am satisfied that the bees of the colony sting the workers and queen confined in the cage through the wire. Some beekeepers hive tried to shorten the period of queenlessness of the colony by putting the mailing cage into the hive before removing the old queen. Their theory is that the bees will get acquainted with the queen, and she can, therefore, be released as soon as the old queen is removed. In practice, however, this failed to work. All workers and frequently the queen are killed in the cage, or else killed as soon as released from the cage. It was found necessary to leave the cage in the hive from two to four days after removing the old queen. Therefore, no time was saved but many queens lost. In the push-in cage, however, there is not this loss from the stinging through the screen, probably because the queen naturally stays on the comb and out of reach of the angry workers of the colony.

There does not seem to be this loss when the round Doolittle cage is used, possibly for the reason that the queen stays near the edge nearest the comb. When a Doolittle cage is used with a queen-excluder inside the cage and the entire canal three or four inches long filled with candy, a large percentage can be successfully introduced. This lacks, however, the one important feature, that the queen can not lay while in the cage, and when releases is not received as readily as a laying queen.

Chapter XXVIII. Disposing of Nuclei at Close of Season.

When the queen-rearing season is over, it is of course necessary to dispose of the nuclei. In warm climates, part may be kept over until spring if they are strong in bees and rich in stores. In case one has extra queens, this is an excellent plan since there is always a great demand for queens in the early spring before any can be reared. This demand is caused by the fact that many colonies come out queenless, and if queens can be procured the colonies would be saved.

I find it profitable to winter our nuclei even as far north as Vincennes. Two twin nuclei, with standard Jumbo size of frames, are placed in a case and packed in sawdust much after the plan used in packing full colonies. In case the nuclei are well supplied with honey and have enough bees to fill the hives nicely, they winter as well as stronger colonies that have more room. If the honey producer can winter a few queens in this manner, he will find that they come in handy the next spring in giving them to queenless colonies and replacing queens that are failing. In requeening such colonies, the whole nucleus (queen, combs and bees) is set over the one to be requeened and united by the newspaper method, for this nucleus will not be needed for queen-rearing since it will probably be a number of weeks before queen-rearing can be started.

However, a large percentage of nuclei must be disposed of as they queens in them will be needed for colonies. An excellent way to accomplish this is to gather a number of nuclei in one place in the center where the group formerly stood, then in the midst of them place a hive with two or three frames of honey. Next, remove the frames from the nucleus hives and brush off all of the bees in front of the hive. Some go in, and others go back to their old location. Finding their nucleus gone they circle around till they comb back to the hive where the bees are fanning when they join these, and as all begin to fan they call the bees in the air to them. After this more nuclei may be emptied and all the bees will at once go into the hive. It is a good plan to have a queen tarpon the entrance in order to catch any virgin that is apt to be overlooked. I usually put in about three pounds of bees. Then I go to a colony that is week and dump the bees in front of the hive. They immediately run into the entrance. In this way weak colonies can be brought up to the required strength. If one wishes, he can put a large number of bees together and give them a laying queen, forming a new colony. I believe it is better to strengthen weak colonies, for we can usually find a few of such.

Now the question arises, "will not the bees go back to their old location the next time they take a flight?" No, they do not, for when they set up fanning as they run into the hive they seem to put themselves into the condition of a new swarm and they will stay anywhere they are placed. I have never seen a single bee return to the old stand after it had once joined the new hive. Another peculiar thing in this connection is the fact that bees never fight against the colony with which they are united and never kill the queen. They seem so demoralized that they are willing to accept things as they find them. In uniting them in this manner it is well to make the colony extra strong, for many of these nucleus bees are old and will die off before spring. This plan of disposing of the nucleus bees has been of much value to me, for it has enabled me to build up weak colonies to good strength and winter them over, where otherwise it would have been necessary to unite these weak colonies to save them, thus reducing the number of colonies.

Packing Cases for Nuclei.

Nucleus packing case.

Many honey producers use extra hive bodies or extracting supers for nuclei; but some will undoubtedly find it more profitable to make or buy a special nucleus hive. In using an extra hive body it is necessary to have an extra bottom-board and cover, so there is little if anything saved in using them. A twin nucleus hive, with room for two frames and a division-board in each side, is a splendid equipment and is hard to beat. It is a source of great satisfaction to have on hand throughout the season a number of queens to use in case of emergency. This is especially true in the spring, for in a large apiary there are sure to be at lest a few colonies which come through the winter queenless. If the beekeeper has an extra queen wintered in a nucleus, this can be united with the queenless colony and it will build up ready for the honey flow in fine style.

In wintering the nuclei in cases if the opening in the winter case were directly in front of the nucleus entrance, the two entrances would be so close together as to cause drifting. We, therefore, make a tunnel one-half inch inside measurement, which is placed at right angles with the nucleus. In this manner the bees fly out at the side instead of in front. The tunnel is whittled off round at the end, and a round hole is bored in the winter case. That the bees may easily find the entrance, the end of this tunnel is painted black. For packing, sawdust is used, four inches on bottom, and six inches on top and sides. It is important to use sawdust which has been kept under cover for one summer which is perfectly dry. Green sawdust in our experience is no better than no packing at all.

Comb made from Airco foundation drawn above an excluder.

After the nuclei are united, the combs must be taken care of to prevent their being destroyed by the wax worm. If the combs are allowed to freeze a few times, they are free from that pest until spring, or if they can be kept in a cool place during the winter they are safe. Do not, however, store them in a warm place such as a basement or attic without first fumigating them. A good way to fumigate is to put them into a close room and burn sulphur. This must be done several times, as the sulphur does not seem to destroy the eggs of the moth. Bisulphide of carbon is more effective, but it is very explosive and is dangerous to use in a room. A good method to employ in using it is to stack the hive bodies up outside, putting a thin super cover on both the top and bottom of the pile. The top hive body should be empty to make room for a dish of the bisulphide of carbon. Pour half a pint into the dish and let it remain till it evaporates. To make doubly certain, another application should be made in two weeks. It is a good plan to examine the combs at times to be sure that no wax worms are present, for it is rather depressing to one's spirits to lift off a hive body in the spring and find everything a mass of webs and all combs destroyed.

Chapter XXX. Supplementary Topics.

Queen-Rearing for the Honey Producer.

Although I have now described the method of queen-rearing as practiced in a commercial way, the primary object of this book is to help the honey producer in rearing his own queens, for if a beekeeper expects to get the most out of his bees he must surely look after them carefully. Let us now adapt parts of the queen-rearing methods just described to the special use of the honey producer and recommend such changes as will best suit his needs. Let us consider that you are a honey producer operating two hundred colonies and upward. How far will the methods just described apply?

In the first place, I feel certain that the use of the swarm box, dipped cells and the manner of having cells finished above the queen-excluder are the very best for the honey producer to adopt. While a little more patience and study are required for its mastery, it certainly pays handsome returns for your time and labor invested. Many successful honey producers all over the world are using similar methods with the best of success. Many who use the grafting method and rear their own queens have informed me that with them it had simply meant the difference between success and failure. However, after the cells are completed and are ripe, they honey producer may branch off from the methods of commercial queen-breeder and adopt those best suited to his needs and circumstances.

Queen-Rearing from Commercial Cell Cups.

Mr. M.T. Pritchard, queen-breeder for the A.I. Root Company, Medina, Ohio, has probably reared more queens than any other man in the world. He uses the Root wooden cell cups into which are pressed the wax cups. He prefers the wooden bases because, he says, they protect the wax cups before and after they are completed by the bees; because they facilitate handling of the cells with their occupants; and last but not least, because he can easily separate the cells when they are finished by the bees. They enable him also to pick out the choicest and best cells from the cell starters and give them to the cell-finishers. These wooden cups are mounted on a bar by using a little pinch of wax. Mr. Pritchard prefers the queenless, broodless method for started the cells. He chooses larvae twelve hours old and endeavors to get the largest larvae of that age. He says, "the larger the better." He determines the age by keeping a record of the time the comb is given to the colony having the breeding queen. As soon as there are eggs in the comb it is an easy matter with him to determine the age of the larvae. He prefers larvae slightly curved but not coiled up.

Mr. Pritchard says there are times when he finds the cell-protectors a great convenience and the wooden cell cup fits the cell-protector better than the dipped cell. Mr. Prichard also believes it better to choose larvae near the center of the comb, as he considers the ones near the bottom of the comb not so well suited for grafting. The high quality of queens turned out by Mr. Pritchard is known the world over. A number of years ago the author visited Medina while queen-rearing was in full blast and learned a number of valuable features of queen-rearing from Mr. Prichard.

Requeening.

Some have preferred to introduce the ripe cell to the colony, while some prefer to introduce the laying queen. For the purpose of improving the stock as well as to replace worn-out queens I have requeened my entire apiary of several hundred colonies every year, and sometimes when I thought I had discovered a better breeder many were requeened two or three times in the year. I find the use of both methods of introduction, sometimes the cell and sometimes the laying queen, is very profitable. I believe it will pay all honey producers to have on hand a number of nuclei in order to have young queens to draw on when circumstances demand. In the main, however, since learning that it is a matter of feed that causes the bees to accept the queen-cells, I am using the cell method more and more.

The principle or law of the bees that a well-fed colony will not destroy cells should have a far-reaching effect with the honey producer, for we can now give ripe cells directly to strong colonies *immediately upon removing the old queen with almost unfailing certainty that they will be accepted.* We must remember that it is hungry bees that tear down cells; well-fed ones do not. When bees are hungry, they themselves tear down cells. When they are well fed they pass the task of cell destruction up to the queen. If she wants to swarm, she does not destroy them. If she does not wish to swarm, she attends to the destruction of the cells herself. True, the workers are doing most of the work; but if you catch the queen and remove her, the cell destruction ceases like magic if the colony has plenty of food. Therefore, if the colony is well fed and if you remove the queen, the cells are allowed to produce queens, and a strange cell will remain untouched the same as one which the bees themselves reared.

Therefore, if the honey producer raises a lot of nice large cells well supplied with royal jelly, he can requeen with very little labor and with small loss of cells. I strongly recommend that requeening be done after the honey flow is well along and the swarming season has passed or nearly passed. If colonies are given cells when they are preparing to swarm they will swarm anyway. They likewise swarm if a laying queen is introduced. I have found that, if a laying queen is introduced to a colony *before* it has any notion of swarming, it does not swarm that year; but, if preparing to swarm, it swarms with the new queen as readily as with the old one. The same thing occurs with a cell; if the cell is given before the colony has any notion of swarming, that colony is effectually prevented from doing so that season. However, except under unusually conditions, requeening with the cell method before swarming season is not advisable, for the absence of a laying queen at this time seriously affects the honey yield of the colony. In case one waits until the honey flow is well along, the colony may be requeened with little loss in strength, for the bees that would have been reared at that time would be too late to help in that harvest.

The method of giving the cell, which I have found entirely satisfactory, is to requeen during a heavy honey flow by simply removing the old queen and putting in a ripe cell. The bees seldom tear this down. However, to do all requeening during such favorable times is not possible. Just as the honey flow closes and the supers are off, you can remove the old queen in the evening and give the colony a heavy feed. Next morning give it a ripe cell, which the bees invariably accept. We must remember that in nature bees do not have cells except when they are receiving food in abundance, and we must duplicate these conditions if we hope to succeed.

After the cell is given, it is well, in a few days, to look to see if the queen has emerged. This may be determined by the appearance of the cell. If it has a small opening at the lower end, you may be sure the virgin is there. If the beekeeper has little time he can wait for ten days, and if eggs are present he may be sure the queen came from the cell given; but if there are other queen-cells started, it indicates that the cell given was destroyed. In such cases the frames should be removed and all bees shaken off in order that every one of these other cells that are started bay be destroyed. Queens coming from such cells would usually be inferior. As the colony has been without a queen for some time, it would be better to introduce a laying queen if you have some on hand, rather than give another cell, for this would leave it queenless so long that it would become greatly deleted in bees. Furthermore, if the second cell given should be destroyed, the colony would probably run into laying workers and be ruined. If you have no laying queen a cell should be given as previously described, feeding heavily as before. However, in this case it would be well to examine it one or two days after the virgin emerged from the cell, and in case this cell is destroyed give another immediately. We should bear in mind that cells are destroyed in this manner very rarely; but, in order to avoid ruining a colony from queenlessness, it should be carefully watched. However, if the work is properly done few cells will be destroyed.

Uniting Bees.

In the previous pages frequent mention has been made of uniting colonies. By far the easiest and safest method is what is known as the "newspaper method." The inventor of this system was Dr. Miller, and it is one of the many splendid things that he has given to beekeepers.

To unite two colonies, place two thicknesses of common newspaper over one colony on the brood-frames, set the other colony that is to be united on top and put on the cover. In a day or two the bees will gnaw away the paper and become acquainted. As they come in contact with each other, a few bees at a time, there is no fighting whatever. The bees in the upper hive body seem to realize that they have been moved, for they mark their new location and do not return to the old one. In very hot weather it is sometimes advisable to punch a hole in the paper with a common lead pencil to prevent suffocation.

Uniting Laying Worker Colonies.

If one has a colony containing laying workers to which he does not wish to introduce a queen, it can be united with any other colony by the newspaper method and the queen will not be injured in any way. The colonies should not be disturbed for a week or more, after which the workers will cease to lay. When this occurs, the hive body which originally contained the laying workers may be set on the bottom board of the present stand and the queen left with them, while the other hive body containing most of the brood may be moved to a new location and a queen introduced.

Making Increase.

There are many methods of making increase. If a large increase is desired, one should begin early in the season. For this, a good way is to proceed in the same manner as described in the first part of this book for forming nuclei, but it is better to give the new colony two frames of brood and bees instead of one. If done early in the season and a honey flow follows, a two-frame nucleus should build up into a good strong colony before winter. If no honey flow is on, they can be built up by feeding sugar syrup. The remainder of the hive is to be filled out with comb or full sheets of foundation. This method should be used only when a large number of colonies are to be made from a few. Where a limited increase is desired the method I have used for a number of years has given such perfect satisfaction that I recommend it in preference to any other.

It is as follows: Take the colony to be divided and set it temporarily to one side. On the stand it has occupied place an empty hive. Take out from the colony one frame of brood with the queen and place it in the empty hive. Then move the colony from which the queen and frame were taken to a new stand and introduce a queen to it or give it a ripe queen-cell, as previously explained. Fill out the hive that remains on the old stand with combs or foundation. The field bees from the colony just moved will return to their old location and build this colony up at a rapid rate. A queen is easily introduced to the moved colony, as the old bees that cause the trouble have gone back "home to mother's." Not enough bees will leave the moved colony, however, to injure the uncapped brood. If only sheets of foundation are used in the new colony it is better to give two frames of brood; but, if drawn combs are used, one frame is sufficient.

Mending Damaged Combs.

How often every beekeeper has dreamed of having perfect combs built clear to the bottom-bar, one hundred per cent worker-cells! All who have such combs please rise. Well, I shall remain seated with the rest. Even when by carefully wired full sheets of foundation we get fairly good combs, in time they get rounded off at the corners and later are drawn out into drone comb. Then mice get in and make holes in the combs. Wax moths do their work, and, as years go by, our combs become more and more filled with drone-cells.

Now, if we take a little pains we can have our combs continually improving and the drone-cells gradually diminishing. It is known that nuclei or weak colonies build worker comb only. Therefore, when we discover a damaged comb, let us put it into a nucleus for repair. If it contains drone comb, cut out the drone cells, and the bees in the nucleus will build worker comb in its stead.

Then in order to get the combs built to the bottom bar, draw the nails in that bar so they will be about ¾" of an inch lower. The bees will build comb to within one-half inch of it, the nails may be driven back into place thereby bringing the bottom-bar up snugly against the bottom of the comb, and you will have a perfect comb as the result.

If it is desired to mend combs when no honey is coming in, the bees will do excellent work at comb-mending by having their division-board feeder kept filled with sugar syrup or honey. Now it is unnecessary to have regular nucleus hives to mend combs, but any weak colony will do it if only two or three combs are given them at one time.

A "mouse-eaten" comb, with bottom bar lowered ready to be mended.

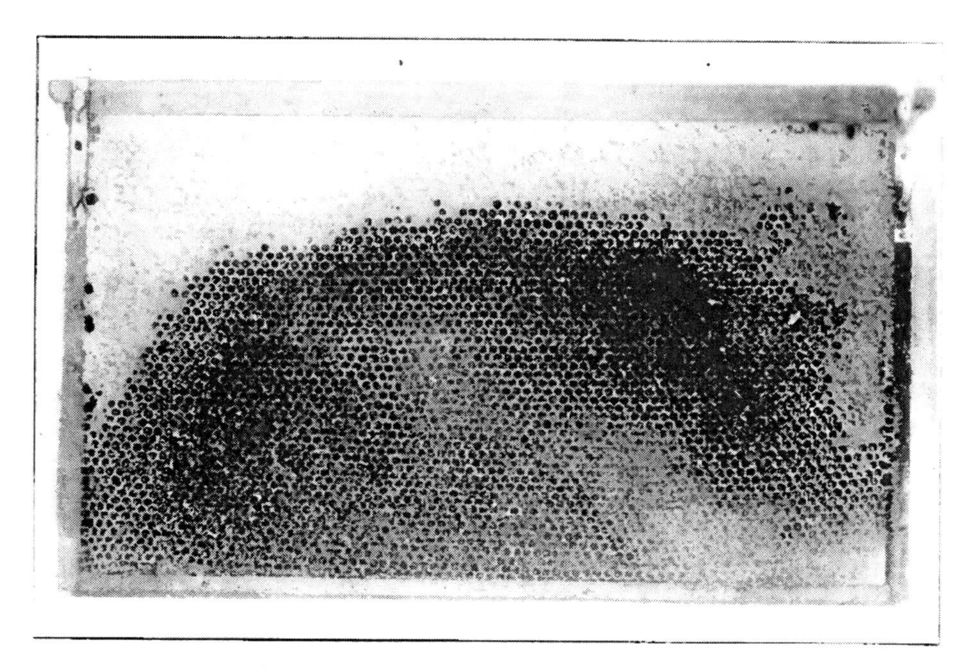

The same comb after mending.

It is the usual rule, where combs are mutilated or imperfect, to cut them out, melt them up and convert into wax. This, of course involved work and expense for foundation and perhaps new frames. Where frames are good and combs, except for holes (for all drone comb should be cut out), are otherwise good they can be repaired by the bees in the manner just described and save expense.

Chapter XXXI. Requeening Colonies About to Swarm.

What is said about giving a cell to a colony preparing to swarm holds good to introducing a laying queen; that is if a queen is introduced to a colony that is in the swarming notion, it will swarm just the same, the newly introduced queen going out with the bees. It is generally conceded that a colony headed by a young queen, one reared the current year, will not swarm.

Now, many amateurs, and some experienced beekeepers, too, have sought to cure the swarming fever by introducing a young queen. Failure is usually the result. A beeman, more progressive than his neighbor beekeeper, in order to get ahead of him, bought a queen from a breeder some distance away. He was very proud of the queen and rubbed it in on his neighbor, telling him he could not buy this queen for $10. It so happened that he introduced her to a colony preparing to swarm. He stopped his beekeeping neighbor and led him out to the hive to show him his fine new yellow queen. Before they got there, they met a large swarm of bees going to parts unknown. When they looked into the hive, there were a lot of queen-cells and few bees! A swarm and that new queen had taken to the woods. It was now his neighbor's turn. He said, "Say, I can now buy that queen for ten dollars can't I?"

However if a young laying queen is introduced to a colony before the colony has the swarming notion, my experience is that when the colony is run for comb honey it will very rarely swarm that season, and when run for extracted honey I have never had one swarm when requeened in this manner.

Hive Body Used as Nucleus.

A very popular method of requeening under certain circumstances is by using one of the brood-chambers where colonies are run two stories high. Certain localities have no early honey flow. The bees build up in the spring, go "over the peak of brood-rearing" and are on the down grade before the honey flow comes on. This is the case in many alfalfa districts.

A splendid method under such circumstances is the following: We shall suppose the colonies are in two-story brood-chambers. As the hives are getting well filled with bees and brood, remove one story, place it on a bottom-board, give it a cover, and put it close beside the hive body from which it was taken. To the queenless part give a rip queen-cell. This will in time build up to a strong colony. When the honey flow begins, kill the old queen and unite these two colonies by the newspaper method described further on. In this manner the colony has been requeened, and as there is a young queen in the hive, there is little danger of swarming.

By this system, the colony has the brood reared from two queens, and becomes a stronger colony than was possible to get from one queen. This same method can be used to advantage in the eastern states where there are a clover flow and a fall flow. In that case, the upper hive body is raised up and supers added between the two hive bodies during the clover flow. When most of the brood in the upper hive have emerged, it is set on a new stand at the side of the parent colony with enough bees to make a small colony. A cell is given in the regular way. This colony is allowed to build up until the fall flow opens, when it is united with the original colony after killing the old queen. By this method of requeening, the colony has been materially strengthened instead of weakened. In other words it has had *two laying queens* for some time instead of *no queen at all* as is the case where the old queen is removed and a cell given.

In case the double brood-chamber is used, one can be employed as a nucleus at a little outlay. All that is necessary is some extra bottom-boards and covers. When one makes a practice of introducing the queen-cell directly to the colony, I believe even then it is a splendid idea to have on hand a small number of nuclei. Laying queens can be kept in these to be used when needed. In working with the bees you occasionally run across a colony that needs a better queen, and if you have some laying queens in reserve in the nuclei, they can be drawn on when needed. Should a cell that you have given to a colony fail to provide a queen for it, a laying queen can be given thus preventing the weakening of the colony from lack of one over too great a period.

Transcriber's picture of a Bottom Board Feeder (the original picture was not clear)

If we use a large brood-chamber, we have to do little feeding, as a rule; but there are times when the honey crop fails, or we have been a little too enthusiastic in making increase, so that we find our colonies deficient in stores, with winter not far away. All beekeepers have been caught in such a predicament at one time or another. The feeder I prefer is made by nailing a strip across the bottom-boards in the deep side about two inches back of the entrance. In fact, when order the regular bottom-board I order an extra piece like the one used for the back cleat on the deep entrance. If this is not going to be used for some time, the bees will stop up all the cracks and make it water tight. If it is to be used at once, pour melted wax or paraffin along the cracks until it is tight. This feeder costs only three or four cents and is always there and never in the way. Moreover, it does not interfere in any way with the ventilation. If a driving rain comes up, this cleat keeps the rain from beating in. If you find a colony that needs feeding, all that is necessary is to slide the hive ahead about two inches on the bottom board and pour the syrup in by allowing it to flow against the back of the hive. It will thus spread out and run down into the bottom-board feeder. A funnel can be used if preferred. This bottom-board feeder holds about ten pounds of syrup. Of course, the hive must be level to prevent the syrup from running out. If a colony needs heavy feeding, the bees may be fed three or four evenings just at dusk, and before morning they will have all of the feed cleaned up away from robbers. This is a very good feeder for stimulating during queen-rearing also. If used every day for this purpose, it is best to have a thin board about two inches wide and the length of the bottom-board across the end as a sort of lid. Then, when feeding, this cover is raised and syrup poured in.

Preparing the feed for winter.

There are several ways of preparing syrup for the bees. Some give it warm, and some recommend cold syrup. I endeavor to feed early in the fall before the weather is too cold. A large tub or boiler is a good thing in which to mix the syrup. Pour in cold water till it is about one-third full. Then add granulated sugar and keep stirring until there is a saturated solution. After you have added so much sugar that it settles to the bottom about two inches thick and does not dissolve, no matter how much stirring you do, the syrup is right. Let it stand a few minutes. It will clear up and you have a nice, clear, smooth syrup. Now, when this is given to the bees they invert it and there is no danger of it's granulating in the hive. If you should follow this process and have the water warm, it would take up too much sugar so that when it became cold crystals would form and it would granulate in the combs. Another objection to the warm syrup is that, when the bees get in it, they become covered with sugar as soon as they dry off, and many are lost in that way. Of course, if warm syrup is given and you have the proportions exactly right, it is a good method but it requires much more work than the cold method just described. No measuring is necessary and water or sugar may be added at will and by stirring, the syrup can be made just right.

We should so conduct our apiary that feeding for winter is rarely necessary. Let us first assign to the bees the task of laying up honey for themselves. Then, after they have done this so that they have an abundance, we may consider that all they make above that amount rightfully belongs to us; but let us bet sure they have enough before we dip in. Sometimes in spite of care, however, the bees will need feeding. In case we have made a large increase or have had a late swarm or total failure of the honey flow, a number of well-filled combs of honey on hand for this purpose is a splendid policy. When a colony needs feeding nothing is better than combs of honey.

Chapter XXXIII. Requeening to Cure European Foul Brood.

It is now generally acknowledges that European foul brood is a disease of weak colonies and inferior or black bees. Many extensive honey producers testify to the fact that, if all colonies are headed by vigorous young, Italian queens and the colony kept strong, so far as European foul brood as a menace is concerned, they can forget it. During the winter of 1918-19 I was employed by the Government to do extension work in the Apiculture Department, my territory being California. During this time I had a rare opportunity to study bee disease, especially European foul brood. The conditions in California are favorable for the development of this disease. I found the statement made above concerning the cure of European foul brood to be correct. In many cases this disease threatened to put the beekeeper out of business until he began to rear queens from good Italian stock and requeen the entire yard. Then, in addition the colonies were given a large brood-nest, usually two nine-frame hive bodies and an abundance of stores. In such cases European foul brood ceased to be a menace. To be sure, it kept the beekeeper on the alert to see that no colony became weakened or had a poor queen. Now as we well know, strong colonies are the profitable ones, so it is evident that European foul brood makes a better beekeeper out of the one that has it. For this reason it is commonly known as a blessing in disguise. We admit that, when this disease first strikes an apiary and causes havoc before the beekeeper can form his defense, the incognito of the blessing is complete!

The method of rearing queens to eradicate this disease is the same as already given, but is better to introduce the ripe cell to the colony affected. Under no circumstances do the bees "clean house" as well as when they have a virgin queen. They clean and polish the cells to make ready for her to deposit eggs. Being without a laying queen they have very little brood to attend to, so they seem to devote all their energies to cleaning out the disease.

The important point in the cure of European foul brood seems to be to have a large number of bees in proportion to the brood. In any cure now being used this condition should be present. When we give a cell as stated, we reduce the amount of brood that the colony has, by removing the laying queen, thus making the cleaning out easier. When the queen is caged, the same condition is brought about. In some cases cures are effected by putting the brood in an upper story and confining the queen to the lower story with a queen-excluder. The same condition exists, for many of the bees leave the queen so that she slackens up on egg-laying, thus reducing the brood. Many have reported that, by putting a new swarm into a colony affected with European foul brood, an immediate cure was effected. This is the same condition as in the others, a large number of bees in proportion to the brood.

It is noticed that the first brood reared in the colony in the spring is not diseased. This same condition, plenty of bees to clean out the disease, is present. A little later in the season the disease is at its worst. This is due to the fact that the bees are rearing the maximum amount of brood in proportion to the bees, as the colony is rearing brood to the fullest capacity, and the old bees are rapidly dying off. If the colony survives and is built up strong, the disease disappears. This is due to the fact that the queen has reached her capacity in egg-production, so the number of bees in proportion to brood is greatly increased. However, if the colony is very weak, it is not a good plan to try to build it up. Better unite it with another colony. If European foul brood is very prevalent in an apiary, I would requeen the entire yard every year until the disease is stamped out. Then every colony that is not strong should be requeened, and if disease shows in any colony it should be requeened.

American Foul Brood.

Let it be understood that all which has been said about requeening to eliminate disease *does not* apply to American foul brood. As this book is a treatise on queen-rearing; we need not discuss American foul brood but only touch on one or two points. American foul brood cannot be cured by requeening. At the present writing, the "shaking treatment" is the only cure advocated. In Dr. E.F. Phillip's excellent book, "Beekeeping," (page 404) the disease and its treatment are described in detail. Many beekeepers wish to requeen colonies having American foul brood-not to cure the disease but to replace the queen on account of her age, for a good young queen is needed in the colony after losing all its brood and having to build up on foundation. The question is very frequently asked, "When is the best time to requeen a colony affected with American foul brood-before shaking or after?" I recommend that they be requeened after shaking; for, if a vigorous queen is introduced and allowed to lay heavily before shaking, she is liable to injured when this is done from the fact that her egg-laying is suddenly stopped, in the same manner as a queen laying heavily is injured by being placed in a mailing cage. Many have reported that queens that were very prolific before the disease was treated, were worthless after the colony had been shaken for its cure. Another reason is that it is not a good policy to open a colony affected with American foul brood any more than is absolutely necessary, on account of the danger of spreading the disease through robbing. After the colony has been treated and has several frames of brood, it is a good time to requeen. If the Push-in cage is used it will be necessary to get an old black comb from some other colony on which to use it, care being taken that it is from a healthy colony.

Chapter XXXIV. Finishing Cells in Queenless Colonies.

Sometimes, conditions are not right for building up colonies sufficiently strong to do the best work at finishing cells above an excluder in a queen-right colony. Nothing but failure will result in attempting to get good cells finished above an excluder if the colony is not *extra strong.* If these extra-strong colonies cannot be obtained, good results may be secured by giving the cells to a colony of medium strength made queenless. To use this method it is first necessary to get the cells accepted, as explained. Then take one bar of cells, go to the colony you wish to use as a finisher and remove the queen, at the same time giving a bar of cells. The bees will go right ahead and finish them in good shape.

The question comes up whether this colony should be queenless for a few hours before giving the bar of cells. It seems to make no difference, as the bees take right hold and go to work on the cells the moment their queen is removed. They seem to do this through their habit of completing a job that has been started. As the queen is not there to protest, the cells are finished. This colony may be used to build a second bar of cells, and even a third, but care must be taken to see that no cells of its own are allowed to furnish a queen.

Finishing Cells While Requeening.

A splendid plan for finishing is while requeening. Go to a colony that is to be requeened. Kill the queen, give the colony a bar of cells that have been started by the swarm box or queenless, broodless method, and at the same opening of the hive introduce a laying queen with the Push-in cage. In four days, by the time the queen is accepted, when you go to take out the introducing cage the cells will be sealed. They are taken out and given to some other queenless colony for incubation, or put over an excluder above a strong colony until ripe and ready to be given to nuclei. In this way you lose no time, for the cells are finished while you are introducing a queen. I find this very convenient late in the season after a honey flow when the colonies are not so strong as they are earlier in the season. As a rule, ten or twelve cells are enough for the bees to finish and do the best work. I never use this method exclusively but do so occasionally in conjunction with the method of finishing above an excluder.

Chapter XXXV. Cell Building During a Heavy Honey Flow.

All are agreed that the ideal time for cell-building is during a light honey flow. However, we have to take the weather and the honey flow as they come, and we seldom have a light honey flow for any great length of time. When the honey flow is beginning conditions are idea for a time; then, as it increases until it becomes a heavy flow, the kind that we all like for honey production, a number of elements interfere with queen-rearing, that much be overcome if we would succeed.

One of these is the scarcity of larvae of the grafting age, due to the fact that the bees are gather nectar so rapidly that the breeding queen is cramped for room to lay eggs. If the case is not too bad, an empty comb inserted in the middle of the brood-nest of the breeding queen will keep her on the job of egg-laying. If the honey flow is heavy, however, this will do little if any good, for the bees at once flood the new comb with nectar. It is their natural instinct, when an abundance of nectar is in the blossoms, to gather it and let everything else go, since other things, such as brood-rearing, cell-building and propolizing, can wait until the rush of harvest is over. Now this rush of nectar into the hive is caused by the field bees; so, if we wish to stop the supply in the colony containing our breeding queen, all that is necessary is to remove the field bees from the colony. This me be easily accomplished by moving the colony out to a new location. The workers, upon returning from the field, will go back to their old location, and if a hive is close to the former position of their own home, will enter without hindrance and put the fruits of their efforts into that hive instead of the one they formerly inhabited. The hive containing the breeding queen will, in consequence, receive very little nectar for several days. In the meanwhile, young bees will emerge, thereby making room for the queen to lay, and brood-rearing can go forward apace. If the honey flow continues, it may be necessary to move the hive several times to keep the excess of nectar from coming int.

Another difficulty, caused by a heavy flow, is found in the cell-finishing colony. Wax will be built all over the cells, sometimes completely covering them. In this case, since we want all the bees possible in the finishing colony, it is not advisable to move the hive as was done with that of the breeding queen. The remedy in this case is to remove all combs as fast as they are filled with honey, being careful not to take way any brood since that would weaken it. In the place of the combs removed, give empty combs. Foundation does not answer, for they will feel crowded for storage room and continue to build comb over the cells. If the bees still build comb around the cells, it is evident that they are yet crowded and an extra story of empty combs should be placed on top. Some extra lifting is necessitated to get at the cells but the results are well worth the work it involves.

Another difficulty arising from a heavy flow is found in the nuclei. This is not so serious as the former, and the only inconvenience is that the combs get crowed with honey and bulged at the tops so that they are removed with difficulty. If only one comb with a division-board feeder is used, the bees will get crowded and go over the division-board and build comb in the empty space. These small bits of comb when removed may be thrown into the solar wax-extractor and if much honey is placed in the newly built combs, it goes well with hot biscuits and is helped along with a glass of Jersey cream! At least, that rule holds good in this locality. IF much of this honey is found, cut it out, put it into tin buckets and tell the neighbors about it. It always moves off at a good price. This condition in the nuclei is remedied by giving empty combs or foundation. I usually prefer to leave one comb with them and give them a frame containing foundation. If the flow continues, the heavy comb is taken away and another frame with foundation is given.

Cell-Building at the Close of the Honey Flow.

By far the most difficult period of queen-rearing is at the close of the honey flow. The flow has restricted brood-rearing, and the fielders have worked themselves to death, so that the colonies are losing strength and the proportion of young bees is small-two serious conditions in queen-rearing. The feeders must be brought into use on the cell builders and on the hives containing bees for the swarm box. Search must be made through other colonies to find frames containing brood, which will be scarce for the reason that all have retrenched in brood-rearing. Some colonies have more brood than others. All combs not containing brood should be taken out and replaced with brood. If this is done after an early honey flow, such as white or sweet clover, cell building will go on nearly as well as at any other time.

After a fall flow, to keep the bees interested in the queen business is much more difficult. Finishing cells in a colony made queenless may be necessary. In case a few flowers, such as asters or goldenrod that the bees may work on, still bloom the queen-rearing season may be prolonged. If brood can be secured good cells can be built, but cold weather and scarcity of drones make queen-mating difficult.

It is a very simple matter to rear queens that are just "queens"; but to rear the very best, those that are long-lived and prolific, and to do this with uniformity under changing seasons and weather, require not only skill and experience but eternal vigilance which is the price of success. Bees observe the change in the nectar secretion much more readily than the beekeeper. From all appearances the honey flow is at its height, with no indication of slacking up as far as the beekeeper can see, but the worker bees may be seen astraddle of the drones, riding them to destruction. The workers see the slackening of the flow. As the only opportunity that the drones might have had is past and they will have outlived their usefulness before any more queens are reared, the bees seem to reason, "Why keep these drones around to eat up the profits?" so the poor fellows are driven out to starve. When the apiarist sees this condition he should put on the feeders, or a poor batch of cells will be the result.

Riding them to destruction.

It is the easiest matter imaginable to rear poor queens. I have seen queens in every stage of size and quality from a worker up to the very finest. Upon several occasions I observed, in nuclei, virgin queens, if we may call them queens, that were no larger than workers. The only way I could tell that they had a touch of queen make-up in their nature was the shape of their abdomen, which was more pointed than a worker's, and the color was little different, showing more yellow. The first one of these I found was a puzzler to me. I had taken out a laying queen and introduced a cell in the regular way. When I looked three days later the cell was torn down, and then it was that I discovered this pygmy queen. I left her to see what would become of her. She disappeared about mating time, as did the few others that were discovered later. Now this pygmy queen emerged from a regular worker-cell, as no sign of a queen-cell or an enlarged worker cell could be seen.

I believe the pygmy queen was reared in this way: When the laying queen was removed the bees began to feed some of the larvae profusely with royal jelly preparatory to rearing queens. When the queen-cell was given and the virgin emerged, these larvae that were given the extra feed or royal jelly were not destroyed but were left to go on and develop as workers; but, as they had received more of the queen food than a well regulated worker should, they took on a slight character of a queen. They have just enough queen nature in them to object to the presence of a queen-cell. Now why was this queen so small? Simply from the fact that she received such a scant supply of royal jelly.

I mention this occurrence to show that all grades of queens can be reared, with no distinct line between a worker and a queen. If the larvae are not supplied with food in abundance, inferior queens result. Then there are other grades of queens, a little larger than the pygmies, which emerge from a queen-cell, but they have been skimped in their food supply. These will be missing at mating time. Some are killed by the workers, that seem to realize the queen is worthless. Queens a little larger can be reared, and the percentage that is missing at mating time is large.

Those that do mate and lay are very inferior, laying very sparingly. They are usually superseded soon; but, if it is late in the season, they die in the winter and the colony comes out in the spring with laying workers. Then there are queens a grade higher that are fair layers. Next, there are good layer, and so on up the line until we get to the very best, that will keep a large brood-chamber packed full of bees and produce a colony that makes several hundred pounds of surplus honey above the average. These are the queens that bring you the profit and the ones you can raise if you play the game according to the rules laid down by the bees.

Chapter XXXVII. Drones.

All recognize the fact that the drone has as much to contribute toward keeping up the stock as the queen. Placing drone combs in colonies containing the best queens, has been advocated to rear drones. If this is done in a large way and drone comb given to a large number of colonies, good results may be obtained. I believe a much better plan, however, is to requeen the entire yard systematically and restrict the drones as much as possible by using full sheets of foundation. Enough drones are reared, and no colony will be injured, as would be the case if a lot of drone combs were allowed in a number of colonies.

Of course, if you are able to keep all drone comb out of the colonies some provision would have to be made for rearing drones, but I have net to see a colony that could not rear a few. Many say that, if they do have all pure drones, it would do little good as their neighbors have hybrids and blacks. There are many times when this would prevent pure mating of queens, but in many cases it would not. We are finding out that queens do not go so far to mate as was formerly supposed. A number have reported seeing queens mate within a very few feet of their hives. I have witnessed the same thing. If there are no bees nearer than a quarter of a mile, and if you have all pure drones in your own yard, my experience would prompt me to say that you will have very little mismating.

As is the case of American foul brood, a lot is laid to the neighbors that rightfully belongs to yourself. I have had several amusing experiences concerning the above. One man said he could not get rid of American foul brood because his bees got it from his neighbor a couple miles away; and when I called on that neighbor, he told me the same thing about the beekeeper to whom I first talked. IT was evident that both were spreading infection among their own bees. So it is in the case of drones; see that all drones in your own yard are pure and you will be surprised at the few mismatings which will occur. Of course, it is still more desirable if you can get the co-operation of your neighbor and interest him in better bees. Get him to requeen by rearing his own, or you can rear them and sell to him.

Chapter XXXVIII. More Than One Queen in a Hive.

Normally, bees permit but one queen in a hive. Frequently, however, during supersedure the old queen remains for some time after her daughter is mated and laying. In such cases the old queen is so incapacitated that the bees and young queen do not seem to recognize her as a queen, and pay no attention to her. These old queens may be dropped into any queenless colony and are generally accepted. They may be utilized in this manner to carry a colony along until it can be requeened with a good queen.

Sometimes a freak case seems to violate all general rules. I have known two virgins to emerge in a hive and be the best of friends. They would mate and lay for quite a while; but sooner or later the bees decide this state of affairs is irregular and kill one of them. Sometimes two queens get into the same hive upon returning from their mating flight. They get along together for a while; but after a time the bees decide they cannot serve two masters. Once faction balls one queen, the other faction the other; thus both queens are so badly crippled that they have to be removed. I remember once when one queen had both wings nipped off close to her body and the other queen lost both legs on the same side. The bees were satisfied then and seemed to think that these two queens, since they both had been trimmed, were about equal to one good one.

A number of years ago there was considerable talk about the possibility of having several queens in one hive, and quite a good deal of experimenting along that line was done by many beekeepers. If it were possible to winter over fifty or one hundred queens in one hive and then have them to supply the demand for queens early the next spring, why, it was worth working for. Personally I was much interested. I used up all the old queens I had left from requeening and sacrificed several dozen good young ones. I learned some very interesting things but little of economic importance. I learned it was not difficult to introduce queens to each other, so they would be friends, yes, regular old cronies, always working together, and usually found on the same comb. The only discovery of importance was the fact that it is the *bees* that make plural queens in a hive impossible.

True, the queens usually fight, but they can be introduced to each other. However, the bees will swear allegiance to only one queen and declare war on all others. Once I put twelve queens in a mason jar to see what would happen. They began to fight, so I shook them around until they were tired. Five had been killed. The other seven called an armistice and apparently signed a treaty that was satisfactory to all. They probably laid the trouble to the dead ones entirely. They held out their tongues to each other and always clustered together. Now, I thought, my troubles were over, for I knew I could safely introduce them. I took six frames of emerging brood, brushed off all of the bees, put the frames into a hive and turned loose the seven queens. They staid together on one frame of brood, and as they emerged, they all began to lay. They certainly did furnish a generous supply of eggs, placing many in each cell. Now I reasons that, as the young bees emerged, they would never know how many queens they were supposed to have and would accept the seven all right.

Things went well for about two weeks, when this small colony began to assume the proportions of a real one. Then trouble developed very rapidly. Having arrived at the age of accountability, the bees decided there were too many queens. This multiplicity of stepmothers, was intolerable, so the killed one. In a few days they killed another, then another. I carefully examined them and found them balling another. For nearly a month this weeding process to get down to one queen continued, but they finally did it and saved the youngest and best.

In concluding this subject, I wish to give as my opinion that it is indeed useless to try to get the bees to tolerate more than one queen for any time long enough to do any real good to the colony. The above is given that it may save others costly experimenting. Of course, if one wishes to experiment merely for the fun of the thing, that is a different story.

In one experiment the stings of two virgins were clipped. These virgins could not fight but mercy, how they did wrestle! The bees stood aside merely interested by-standers until one of the virgins began to squeal, when the workers closed in and balled both queens and injured them so it was necessary to replace them.

As stated an old queen is readily accepted. Usually when a queen gets old and lays but few eggs, she is killed by the queen that supersedes her or dies a natural death. I observed one notable exception, however. I requeened a colony yearly for two years and then found an old queen in the hive that had been there for four years. I could tell her by the manner in which her wings were clipped. She had been in the colony, and a young queen was reared to supersede her. This queen was removed when one year old; and another a year later. As far as I could observe, this old queen had ceased laying altogether and was treated with absolute indifference by both bees and queens.

Chapter XXXIX. When to Requeen.

The question is asked many times, "When is the best time to requeen?" The answer depends upon the strength of the colony and upon the location. From the latitude of St. Louis south, when the colony is strong, late requeening is desirable. In fact, the later the better. In such a case no benefit will result to the colony requeened that season, for the queen would lay little if any before spring. Being strong, the colony would have plenty of bees to come through the winter in good condition. The queen would begin laying early in the spring and have the colony on the job for the first honey flow. If the colony is of medium strength requeening is preferable about August first, for then there is time to build up and go into winter quarters with young bees.

Further north, requeening should be done in August, for, on account of the long winter, it is desirable to have as many young bees as possible to carry the colony safely through, as young bees not only live longer but stand confinement better than old ones. In the north, if the colony is not strong it should be requeened in June or July, depending on its strength. In case one wishes to build up a weak colony for a fall flow, the earlier it is requeened in the season the better. Some of the colonies that have done the best work for me in a honey flow from sweet clover in June were requeened in November of the previous year. They were strong when they were re-queened.

In many parts of California, the colonies run down near the close of the season, winter poorly and build up on the first honey flow in the spring, thus failing to get a surplus from that flow. In such cases if the colony were requeened in August and given plenty of stores, the queen would build up the colony to good strength for winter. She would take a rest until January or February and then build the colony up strong, ready to give a surplus from the first flow.

Now from the fact that it has been advised to requeen early in the season, many get the impression that it is not a good plan to requeen late in the season, even if the colony has a poor queen. This is not correct, for a good young queen is always better than an old one. Therefore, late in the season, if you discover any colonies that have inferior queens, by all means replace them with younger ones if they can be procured. If the colony is weak, it is probably better to kill the queen and unite it with another. If the colony is of good strength and a queen is introduced in October or November, she would be in the very best condition to build up that colony to great strength early the following spring.

How Often to Requeen.

A great difference of opinion exists as to how often to requeen. Some say requeen every year; some say, every second year; and some say, requeen only the colonies that have poor queens; while others say, let the bees do it themselves. Circumstances have to determine; but it is safe to say that there should be vastly more requeening than is practiced at present. I believe in most parts of California the honey crop could be doubled if each colony would be requeened every year with good Italian queens reared by the beekeeper himself, especially if a large brood-nest full of stores should be provided.

In many other parts of the United States, the same or similar conditions prevail. In the far north where the queen is idle such a long periods in the winter, possibly it does as well to requeen every two years as to do so every year where the honey season is longer and the queen kept on the job ten months. In this locality I have found it profitable to requeen every year, with an exceptional case where queens of unusual qualities are found which are desired for breeders. I believe a good rule that applies to all localities is to go over all colonies each year and replace every queen that is not the very best. In case a queen has a big brood-nest packed with brood, is large and has all indications of being a splendid, prolific queen, I would leave her another year. One can usually tell by the size and general appearance of a queen whether she is to be trusted to do the best work for another year or not. If there is any doubt about it, err on the safe side and requeen. Another point, that should not be lost sight of in requeening, is that one can continually improve his bees by breeding from the best. We should also keep in mind, if we requeen every year and keep all colonies strong, we shall not have to "cure" European foul brood, because we will prevent it.

Some have advocated that the best breeder is one that lives the longest. Now that depends on the amount of work she has done. I believe the best beekeeper is the man who can get the most eggs out of a queen in the shortest time. If managed properly, the queen should lay the bulk of her eggs the first year. In the north exceptional queens do well the second year; but, if a queen does even fair work the third year, it is evident she was not worked as she should have been during the two years previous.

Sometimes beekeepers believe they have discovered an exceptional queen from the fact that she lived four or five years. Possibly the queen *was* a good one; but I have some doubts about the beekeeper himself, for, if he had given her a chance, she could have done her work in one or two years instead of loafing along for four or five.

Chapter XL. Commercial Queen-Rearing.

The question is frequently asked, "Does commercial queen-rearing pay?" So much depends on the person and locality that it is difficult to answer this question in a satisfactory manner. If one is in a reasonably good honey producing location, he can make much more money at honey production. Much more skill and experience are required to make a success of queen-rearing than to make a success of honey production. However, as we believe every honey producer should learn the queen-rearing business, some will prefer queen-rearing.

The main requirement for a successful commercial queen-breeder is a love for the business. If he has that, no obstacle is too great for him to overcome. The many difficulties in queen-rearing on account of failures in the honey flow and unfavorable weather conditions will cause all but the stoutest hearts to throw up the business in disgust. However, some prefer to engage in the queen business for the enjoyment and satisfaction they get out of it, even if the financial returns are not so great as in honey production. To attend to ordinary queen trade requires the handling of many details, much time must be given to bookkeeping and correspondence; therefore, additional help is required in these lines. On the other hand, the demand for queens of high quality is great and a good queen-breeder should experience no trouble in disposing of his output.

Queen-rearing involves no heavy work, but it is an exceedingly busy job while the season lasts. It is an ideal occupation for women or men who cannot do heavy work, but who are willing to be on the job, early and late during the queen-rearing season. It is not a "get rich quick" occupation and takes a number of years of careful study and experimenting before one can get things lined up and moving smoothly. Considerable expense is connected with it for office help, shipping cages and other supplies, stationary, and for sugar with which to feed the bees in cell-building and swarm-box colonies. The honey crop largely has to be sacrificed, for, as one uses brood to form nuclei, the colonies, thus weakened by the loss of such brood will do well if they build up strong and make enough honey to carry them over to the next season. This holds true in a season when the honey flow is good. In a poor season large quantities of sugar must be purchased to build up the colonies and provide winter stores.

Locality plays an important part. Many localities similar to ours have a succession of light honey flows which are suitable to queen-rearing, but not heavy enough to give a surplus. For instance, our season opens with peach bloom followed by pear, apple, locust and a little tulip tree and tupelo. Then come white clover and alsike, which are very uncertain, and are frequently mixed with a vile honeydew that ruins the honey for market. Sweet clover is next, followed by a short dearth. Sometimes a little honey comes from unknown sources, probably from throroughwort, figwort or ironweed. Then a little climbing milkweed or blue vine is followed by the heaviest flow we have, heartsease. This flow rapidly tapers off into aster and goldenrod. If one is in a locality where there is but one heavy flow he will do much better at honey production, since queen-rearing during a dearth of pasture for any length of time is both difficult and expensive.

Personally, queen rearing is so fascinating and enjoyable that I would prefer to rear queens, even if the financial returns were but one-half as much as from honey production; for, after all, it is the pleasure we get out of life, not the money, that counts. We should follow that occupation which most enjoy.

After all, what is money for? To purchase enjoyment in one form or another. Therefore, if we are getting the enjoyment from our occupation direct, it is the same as money and we save the middleman's profit.

To the commercial honey producer, I can truthfully say I believe there is a bright future for him. More and more we see health experts call attention to the value of honey as a wholesome hygienic food for young and old, for the ill and well. Let us all co-operate to get honey into more homes as a regular diet, at a reasonable price. Especially should we practice better methods of honey production. First, increase our output per colony, and then increase the number of our colonies. With these improvements, there is no reason why beekeeping should not be profitable and, as Dr. Miller used to say, "Just think of the fun I have had."

About the Author

Jay Smith

Jay Smith was an active writer in the bee journals during his day and wrote two of the most loved and used of the queen rearing books. Both are still being followed by many today. He is most famous for this book, *Queen Rearing Simplified*, but *Better Queens* is the culmination of his work in queen rearing.

CPSIA information can be obtained
at www.ICGtesting.com
Printed in the USA
FFOW01n1000201215
19833FF